Disclaimer

The publisher of this book is by no way associated with the National Institute of Standards and Technology (NIST). The NIST did not publish this book. It was published by 50 page publications under the public domain license.

50 Page Publications.

Book Title: Effects of JPEG 2000 Image Compression on 1000ppi Fingerprint Imagery

Book Author: Shahram Orandi; John M. Libert; John D. Grantham; Kenneth Ko; Stephen S. Wood; Jin Chu Wu

Book Abstract: This paper presents the findings of a study conducted to measure the impact of JPEG 2000 compression on fingerprint imagery at various levels of compression. The impact of compression is measured in terms of impact to both Galton and non-Galton based features of a fingerprint by utilizing the professional judgment of trained and seasoned fingerprint examiners. This impact is analyzed by consolidating and quantifying multiple decisions and associating a cost with the different levels of image compression loss incurred during compression. In addition to measuring the perceived visual impact of compression on the aforementioned features of the fingerprint as a result of compression, this paper also examines the impact of compression on the examiner's ability to render identification decisions.

Citation: NIST Interagency/Internal Report (NISTIR) - 7778

Keyword: Fingerprint compression; 1000ppi fingerprint imagery; JPEG 2000 fingerprint compression

Effects of JPEG 2000 Image Compression on
1000 ppi Fingerprint Imagery

NIST Interagency Report 7778

Shahram Orandi, John M. Libert, John D. Grantham,
Kenneth Ko, Stephen S. Wood, Jin Chu Wu

Image Group

Information Access Division

Information Technology Laboratory

National Institute of Standards and Technology

May 25, 2011

ACKNOWLEDGEMENTS

The authors wish to give special thanks to the following individuals and organizations for their support of this work:

- Federal Bureau of Investigation for all their support throughout this study
- T.J. Smith and the LA County Sheriff's Department
- R. Michael McCabe, IDTP
- Margaret Lepley, MITRE
- Tom Hopper, Cogent Systems
- Stephen Meagher, Dactyl ID, LLC
- Ron Smith & Associates

In addition, we appreciate the guidance, support and coordination provided by Michael Garris without whose help this study would not have been possible.

DISCLAIMER

Specific hardware and software products identified in this report were used in order to perform the evaluations described in this document. In no case does identification of any commercial product, trade name, or vendor, imply recommendation or endorsement by the National Institute of Standards and Technology, nor does it imply that the products and equipment identified are necessarily the best available for the purpose.

EXECUTIVE SUMMARY

The criminal justice communities throughout the world exchange fingerprint imagery data primarily in 8-bit gray-scale and at 500 pixels per inch [1] (ppi) or 19.7 pixels per millimeter (ppmm). The Wavelet Scalar Quantization (WSQ) fingerprint image compression algorithm is currently the standard algorithm for the compression of 500 ppi fingerprint imagery. WSQ is a "lossy" compression algorithm. Lossy compression algorithms employ data encoding methods which discard (lose) some of the data in the encoding process in order to achieve an aggressive reduction in the size of the data being compressed. Decompressing the resulting compressed data yields content that, while different from the original, is similar enough to the original that it remains useful for the intended purpose. The WSQ algorithm allows for users of the algorithm to specify how much compression is to be applied to the fingerprint image at the cost of increasingly greater loss in fingerprint image fidelity as the effective compression ratio is increased (see Figure 1 for an example of image degradation from lossy compression). The WSQ Gray-Scale Fingerprint Image Compression Specification [WSQ] provides guidance based on an International Association for Identification (IAI) study [FITZPATRICK] to determine the acceptable amount of fidelity loss due to compression in order for a WSQ encoder and decoder to meet FBI certifications. These certifications are designed to ensure adherence to the WSQ specification and thereby to ensure fidelity and admissibility in courts of law for images that have been processed by such encoders and decoders.

For 1000 ppi (39.4 ppmm) fingerprint imagery, MITRE has developed an informative guidance ("Profile for 1000 ppi Fingerprint Compression" [MTR1]) that widely is recognized as the *de facto* standard guidance for utilizing JPEG 2000 for the compression of fingerprint imagery at 1000 ppi. This document provides an excellent basis for a compression profile for 1000 ppi fingerprint imagery using JPEG 2000, particularly in its specification of software parameters for control and structure of the JPEG 2000 code stream, and its recommendations form the basis for the compression strategy used in this study. The MITRE guidance is informative, however, and makes no attempt to prescribe an optimal compression rate. This study builds on the MITRE work, both to illuminate the effects of compression on fingerprint features used for identification by trained examiners in the context of the MITRE JPEG 2000 profile for 1000 ppi imagery, and also to develop a basis from which a normative guidance regarding application of JPEG 2000 to fingerprint images can be established.

Furthermore, this study extends the work of the IAI with 500 ppi WSQ images to 1000 ppi images compressed using JPEG 2000. Based on the IAI criteria for non-Galton feature loss in formulating its 500 ppi guidance, findings detailed in this report suggest that a compression ratio lower than 15:1 be considered in a normative guidance for application of JPEG 2000 to 1000 ppi images. This is partly due to the experimental observation that the generalized behavior of JPEG 2000 is not as optimal as WSQ for fingerprint imagery. Therefore a less aggressive compression strategy with JPEG 2000 at 1000 ppi is needed to place it on an equal footing with WSQ for fingerprint imagery.

This study examines the effects of increasing compression in terms of image degradation as observed by examiners as well as the effects of increasing compression on the ability of examiners to make their identification decisions. This study confirms that increasing compression does result in progressively greater image degradation. It finds that this degradation impacts some types of images more aggressively than others (e.g., matched rolled pairs are affected more than matched flat pairs). Although the study confirms that, even at the highest compression ratio examined (38:1), the correctness of the examiners' identification decisions is not affected in any statistically significant way, the study does not attempt to measure either the total effort necessary to reach these decisions or the examiners' confidence in those decisions.

---

[1] Resolution values for fingerprint imagery are specified in pixels per inch (ppi) throughout this document. This is based on widely used specification guidelines for such imagery and is accepted as common nomenclature within the industry. SI units for these will be presented only once.

VERSION HISTORY

Date	Activity
05/25/2011	Initial Release

Table of Contents

1. Investigative Goals and Objectives 17
    1.1. Background 17
    1.2. Market drivers 18
2. Materials and Methods 21
    2.1. Compression Algorithm 22
    2.2. Methodology 24
    2.3. Participants 26
3. Analysis 27
    3.1. Observed Image Quality 27
    3.2. Identification Errors 36
4. Results 43
    4.1. Investigative Goal 1: Validate 15:1 target compression ratio 43
    4.2. Investigative Goal 2: Examine image degradation relative to compression ratio 44
    4.3. Investigative Goal 3: Assess impact of compression on identification error rates 45
    4.4. Investigative Goal 4: Examine compression anomalies relative to impression type 47
5. Conclusions 51
6. Future Work 51
References 53
    Publications and Reports 53
    Standards 55
Appendix A. IAI WSQ Compression / Decompression Study Summary 57
Appendix B. Dataset Makeup 59
Appendix C. Equipment Used for Study 65
Appendix D. Observation Data 67
    Ink Capture Compression Degradation Observations 67
    Live Capture Compression Degradation Observations 68
    Ink Capture Compression Identification Observations 69
    Live Capture Compression Identification Observations 70
Appendix E. Statistical Parameters of Bootstrap Results and Hypothesis Testing 71

LIST OF TABLES

Table 1 - Abbreviations 13
Table 2 - Impression Types Selected for This Study 21
Table 3 - Pairing of Each Original Images to Compressed Counterpart 22
Table 4 - Compression Levels Used in Study 23
Table 5 - Other JPEG 2000 Compressor Settings Used in Study 23
Table 6 - Observation Codes for Compression Degradation Observation 24
Table 7 - Normalized Ratings and Assigned Costs 28
Table 8 - Z-Scores 32
Table 9 - P-Values for 2-Tailed Tests 33
Table 10 - Observation Codes for Determination of Identity 36
Table 11 - Normalization Table for Identification Decision 36
Table 12 - IAI Study Results 57
Table 13 - Key protocol differences between IAI WSQ study and This Study 58
Table 14 - Ink Card Scan Data classification by Impression Type 60
Table 15 - Ink Card Scan Pattern Classification for Single Finger Images by Impression Type 60
Table 16 - Ink Card Scan Pattern Classification for Single Finger Images by Finger (Females) 60
Table 17 - Ink Card Scan Pattern Classification for Single Finger Images by Finger (Males) 60
Table 18 - Live-Scan Data classification by Impression Type 61
Table 19 - Live-Scan Pattern Classification for Single Finger Images by Impression Type 61
Table 20 - Live-Scan Pattern Classification for Single Finger Images by Finger (Females) 61
Table 21 - Live-Scan Pattern Classification for Single Finger Images by Finger (Males) 61
Table 22 - Gender Breakdown for Data 62
Table 23 - Age Breakdown for Data 62
Table 24 - Other Metadata: Height and Weight 62
Table 25 - Other Metadata: Eye Color 62
Table 26 - Image Geometry Data 63
Table 27 - Ink Capture Degradation Results 67
Table 28 - Live Capture Degradation Results 68
Table 29 - Ink Capture Identification Results 69
Table 30 - Live Capture Identification Results 70
Table 31 - Distribution parameters of bootstrap replications and hypothesis tests of differences in degradation between lossless baseline and compressed images (Ink Card Scan) 71
Table 32 - Distribution parameters of bootstrap replications and hypothesis tests of differences in degradation between lossless baseline and compressed images (Digital Live Capture) 72

LIST OF FIGURES

Figure 1 - Example of Fidelity Degradation Due to Lossy Compression (JPEG 2000 at 800:1) 18
Figure 2 - Example of Behavioral Differences between JPEG 2000 and WSQ (15:1 compression) 19
Figure 3 - Split-Screen Presentation of Image Pairs 25
Figure 4 - Trending of Responses with Increasing Compression 27
Figure 5 - Histogram typical of degradation score data replicates output from bootstrap procedure with the normal distribution function fitted to data. 29
Figure 6 - Normal probability plot typical of degradation score data. Normality is indicated by linearity. 30
Figure 7 - Mean compression anomaly measure with 95 % confidence intervals for inked, rolled-to-rolled, fingerprints (match pairs) for compression ratios from 1:1 to 38:1. 34
Figure 8 - Identification error rates for ink card scan rolled to ink card scan flat fingerprint comparisons at the 14 compression levels with confidence limits. 38
Figure 9 - Identification error rates for digital live capture rolled to digital live capture flat fingerprint comparisons at the 14 compression levels with confidence limits. 39
Figure 10 - Identification error rates for ink card scan flat to ink card scan rolled fingerprint comparisons at the 14 compression levels with confidence limits. 39
Figure 11 - Identification error rates for digital live capture flat to digital live capture rolled fingerprint comparisons at the 14 compression levels with confidence limits. 40
Figure 12 - Probability of inconclusive determination vs. compression ratio for 10 comparison scenarios and 14 compression ratios. 41
Figure 13 - Summaries over all comparison scenarios showing least squares fit to the mean probability of inconclusive determination. Slight increasing trend is noted with increasing compression. 42
Figure 14 - Observed Compression Anomalies for Case-1 (Ink Rolled to Ink Rolled) 43
Figure 15 - Z-Score Trends (linear) for Ink Card Scan 44
Figure 16 - Z-Scores Trends (linear) for Digital Live Capture 44
Figure 17 - Comparison of Very High Compression Rates 45
Figure 18 - Z-Scores for Ink Card Scan Cases 47
Figure 19 - Z-Scores for Digital Live Capture Cases 48
Figure 20 - Impression Comparison Examples 59

TERMS AND DEFINITIONS

The abbreviations and acronyms of Table 1 are used in many parts of this document.

Table 1 - Abbreviations

NIST	National Institute of Standards and Technology
JPEG	Joint Photographic Experts Group – ISO/IEC committee developing standards for image compression
IAI	International Association for Identification
FBI	Federal Bureau of Investigation
SIVV	Spectral Image Validation/Verification Metric
NBIS	NIST Biometric Image Software
WSQ	The Wavelet Scalar Quantization algorithm for compression of fingerprint imagery
NGI	Next Generation Identification
IAFIS	Integrated Automated Fingerprint Identification System

ABSTRACT

This paper presents the findings of a study conducted to measure the impact of JPEG 2000 compression on 1000 ppi fingerprint imagery at various levels of compression. The impact of compression to both Galton and non-Galton based features of a fingerprint is measured by utilizing the professional judgment of trained and seasoned fingerprint examiners. This impact is analyzed by consolidating and quantifying multiple decisions and associating a cost with the different levels of image compression loss incurred during compression. In addition to measuring the perceived visual impact of compression on the aforementioned features of the fingerprint as a result of compression, this paper also looks at the impact of compression on the examiner's ability to render identification decisions.

KEYWORDS

Fingerprint compression; 1000 ppi fingerprint imagery; JPEG 2000 fingerprint compression

1. Investigative Goals and Objectives

In July of 2009 NIST in partnership with the Federal Bureau of Investigation (FBI) commenced an investigation on the use of JPEG 2000 [JPEG2K] for compressing fingerprint imagery with the following objectives:

1. **Validate 15:1 target compression ratio:** Validate the 15:1 effective compression ratio target for fingerprint imagery captured at 1000 ppi as defined in the current informative guidance for 1000 ppi [MTR1] using the legacy methodology that formed the 500 ppi guidance [FITZPATRICK].
2. **Examine image degradation relative to compression ratio:** Assess if higher compression ratios result in increased perceived image degradation and note any patterns in degradation relative to compression.
3. **Assess impact of compression on identification error rates:** Assess if higher compression ratios result in increased identification error rates for fingerprint examiners and note any patterns in the error rates.
4. **Examine compression anomalies relative to impression type:** Determine whether any particular fingerprint impression types are more susceptible to compression related anomalies than others at the various examined compression ratios.

While addressing the above objectives, the investigators also set out to expand upon prior work that examined only rolled-single-fingerprint images by also taking into account flat, rolled and slap fingerprint imagery for both inked card scan and digital live scan modalities.

1.1. Background

The criminal justice communities throughout the world exchange fingerprint imagery data primarily in 8-bit gray-scale at 500 pixels per inch (ppi). The Wavelet Scalar Quantization (WSQ) fingerprint image compression algorithm is currently the standard algorithm for the compression of 500 ppi fingerprint imagery. The WSQ standard defines a class of encoders and decoders with sufficient interoperability to ensure that images encoded by one compliant encoder can be decoded by any other compliant decoder.

WSQ is a "lossy" compression algorithm. Lossy compression algorithms employ data encoding methods that discard (lose) some of the data in the encoding process in order to achieve an aggressive reduction in the size of the data being compressed. Decompressing the resulting compressed data yields content that, while different from the original, is similar enough to the original that it remains useful for the intended purpose. *Lossless* compression algorithms on the other hand can produce a compressed image that can be decompressed back to original form with no loss or change to the resulting image. The disadvantage to lossless algorithms is that they produce compressed images that can be many times larger than compressed images produced by lossy algorithms.

The lossy WSQ algorithm allows for users of the algorithm to specify how much compression to apply to the fingerprint image at the cost of increasingly greater loss in fingerprint image fidelity as the effective compression ratio is increased (see Figure 1 for example of image degradation from lossy compression). The WSQ Gray-Scale Fingerprint Image Compression Specification [WSQ] provides guidance for the acceptable amount of fidelity loss due to compression in order for the encoder and decoder to meet FBI certifications for 500 ppi fingerprint imagery. These certifications are designed to ensure adherence to the WSQ specification to ensure fidelity and admissibility in courts of law for images that have been processed by such encoders and decoders.

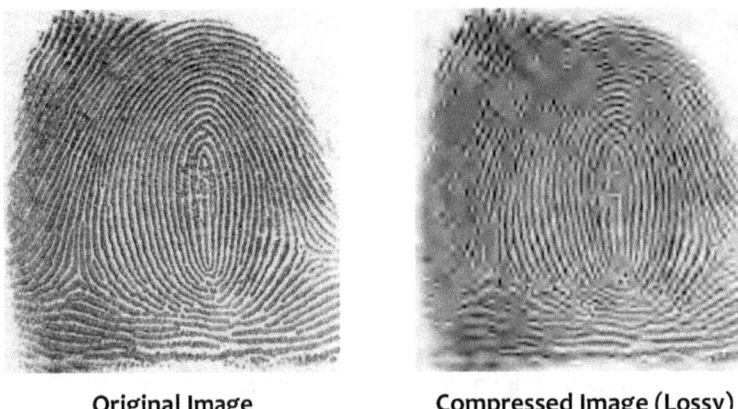

Original Image   Compressed Image (Lossy)

Figure 1 - Example of Fidelity Degradation Due to Lossy Compression (JPEG 2000 at 800:1)

A study conducted by the International Association for Identification (IAI) [FITZPATRICK] established 15:1 as the maximum WSQ compression ratio that would retain acceptable image fidelity. The study used the opinions of expert fingerprint examiners to measure the fidelity loss due to compression. In order to reduce bias due to subjectivity, multiple examiner decisions were used to build consensus. Utilizing examiner opinion does not imply that automated fingerprint matcher performance is not an important criterion in a given biometric system, but it must be noted that if fingerprints are to be admissible as evidence in a court of law their ultimate utility lies in the expert examiner's opinion of the fidelity of those fingerprints.

1.2. Market drivers

In modernizing its environment as part of the Next Generation Identification (NGI) initiative, the FBI seeks to expand its ability to exchange fingerprints at 1000 pixels per inch in an effort to improve upon the capacity of systems in fingerprint identification and verification tasks and meet the FBI mandate to:

- Protect the United States from terrorist attack, foreign intelligence operations and espionage
- Support federal, state, local and international partners in their efforts to prevent or reduce crime and violence
- Upgrade technology to support the FBI's missions

Toward meeting this goal, the FBI seeks to set guidance for the next generation encoders and decoders based on the open JPEG 2000 compression standard in order to ensure interoperability, fidelity and admissibility in courts of law for 1000 ppi images in the criminal justice community.

In support of the FBI, the National Institute of Standards and Technology (NIST) conducted a study to determine an optimal compression approach that follows on the IAI study of WSQ compression for 500 ppi fingerprint imagery, build upon existing guidance for JPEG 2000 compression of fingerprint imagery, and formulate a basis with which a normative compression guidance can be established in the ANSI/NIST standard [AN27]. NIST has an established expertise in evaluating biometric systems and standards, and has been assigned by the USA PATRIOT Act (Public Law 107-56) the responsibility for developing and certifying biometric technology standards. NIST has been supporting biometric standards and evaluation activities for over forty years, starting with fingerprint analysis which began in 1965.

MITRE has developed an informative guidance that is widely recognized as the *de facto* standard guidance for utilizing JPEG 2000 for the compression of 1000 ppi fingerprint imagery in MTR-04B0000022 [MTR1]. While this document provides an excellent basis for a compression profile for 1000 ppi fingerprint imagery using JPEG 2000, some of the informative guidance provided is based loosely on existing accepted guidance for WSQ at 500 ppi.

It should also be noted that unlike WSQ, JPEG 2000 supports both lossy functionality similar to WSQ (see 1.1), as well as a lossless mode. This study examines JPEG 2000 in lossy mode as it has been prescribed in the existing guidance [MTR1].

A preliminary analysis of WSQ behavior vs. JPEG 2000 was conducted early in the present study in which empirical observations were made that indicated significant behavioral differences between the two compression algorithms. These observations reinforced the need to conduct a full study of JPEG 2000 compression at 1000 ppi rather than attempting to incorporate existing informative guidance on a normative basis. This exploratory evaluation of compression algorithm behavior was conducted at 500 ppi due to the 500 ppi limitation of the WSQ's inherent design. In conducting this exploratory evaluation, compression parameters based on the 1000 ppi profile [MTR1] were adjusted for use of the JPEG 2000 compressor operating at 500 ppi.

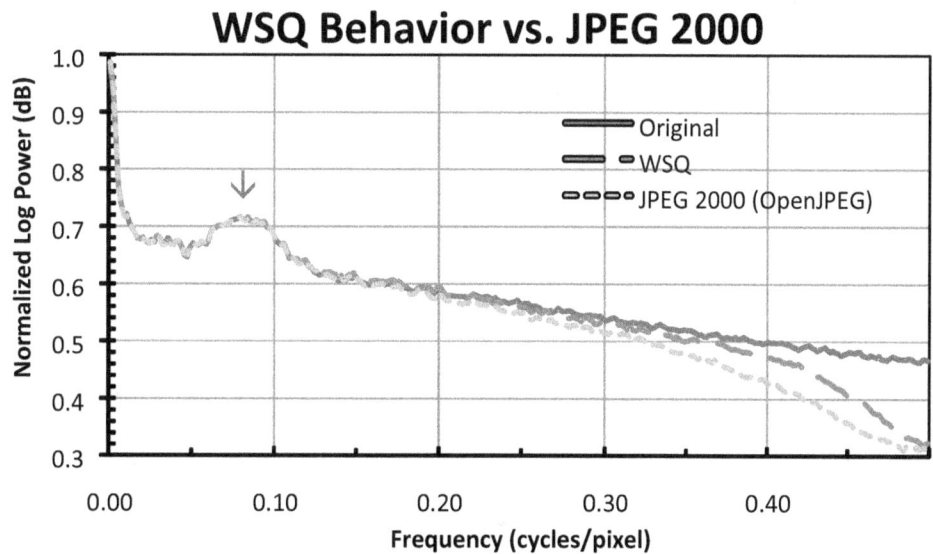

Figure 2 - Example of Behavioral Differences between JPEG 2000 and WSQ (15:1 compression)

The example in Figure 2 shows spectral energy [LIBERT] for a rolled 500 ppi fingerprint image without compression (labeled "Original") and the spectral energies of the corresponding WSQ and JPEG 2000 compressed copies of this image. WSQ maintains near-identical image energy fidelity well into 0.35 cycles/pixel, whereas JPEG 2000 appears to lose image energy at frequencies as low as 0.20 cycles/pixel. Image ridge information is represented in this figure as a peak located at approximately 0.10 cycles/pixel on the image energy spectrum. This demonstrates WSQ outperforms JPEG 2000 by maintaining fidelity over a greater range than JPEG 2000. This is primarily due to the fact that WSQ has been very specifically tuned and designed for its intended purpose of compressing gray-scale fingerprints, whereas JPEG 2000 [2] was designed as a more generalized algorithm for applications beyond fingerprints, such as color photographs.

The empirical evidence observed in this preliminary study demonstrated the fact that the accepted 15:1 guidance for WSQ may not be directly applicable to JPEG 2000 due to systematic differences between the two algorithms, and that this behavioral difference warrants an in-depth study of the JPEG 2000 algorithm and its application in fingerprint compression.

---

[2] JPEG 2000 Part 2 provides for additional wavelet filter customization that may improve its performance for fingerprint image compression. However, these features are not compatible with most JPEG 2000 decoders, so they were not included in the MITRE JPEG 2000 profile, nor were they included in the present data preparation.

2. Materials and Methods

In this experiment a deck of 4,000 1000 ppi images was assembled by sampling randomly from actual operational data supplied to NIST by state and federal partners. The 4,000 image set contained various image impression types as described in Table 2 from a total of 223 unique individuals (100 for ink card scan, and 123 for digital live capture) as described in Appendix B and this set was sampled for both match and non-match pairings. Also, as noted in Table 2, image pairs were selected from both match samples (where the two images were from the same finger/same person) as well as non-match samples of corresponding fingers of the same Galton [GALTON] three-pattern classification (the same pattern of either Arch, Loop or Whorl, but from different individuals and therefore non-matching).

Table 2 - Impression Types Selected for This Study

Case Number	Data Medium	Source (Original) Image	Comparison Image (Compressed, plus Control Case)	Match Pair	Count
1	Scanned Fingerprint Card	Rolled single finger	Rolled single finger	Yes	200
2	Scanned Fingerprint Card	Rolled single finger	Rolled single finger	No	200
3	Scanned Fingerprint Card	Flat single finger	Rolled single finger	Yes	200
4	Scanned Fingerprint Card	Flat single finger	Rolled single finger	No	200
5	Scanned Fingerprint Card	Rolled single finger	Flat single finger	Yes	200
6	Scanned Fingerprint Card	Rolled single finger	Flat single finger	No	200
7	Scanned Fingerprint Card	Flat single finger	Flat single finger	Yes	200
8	Scanned Fingerprint Card	Flat single finger	Flat single finger	No	200
9	Scanned Fingerprint Card	Four finger slap	Four finger slap	Yes	200
10	Scanned Fingerprint Card	Four finger slap	Four finger slap	No	200
11	Digital Live Capture	Rolled single finger	Rolled single finger	Yes	200
12	Digital Live Capture	Rolled single finger	Rolled single finger	No	200
13	Digital Live Capture	Flat single finger	Rolled single finger	Yes	200
14	Digital Live Capture	Flat single finger	Rolled single finger	No	200
15	Digital Live Capture	Rolled single finger	Flat single finger	Yes	200
16	Digital Live Capture	Rolled single finger	Flat single finger	No	200
17	Digital Live Capture	Flat single finger	Flat single finger	Yes	200
18	Digital Live Capture	Flat single finger	Flat single finger	No	200
19	Digital Live Capture	Four finger slap	Four finger slap	Yes	200
20	Digital Live Capture	Four finger slap	Four finger slap	No	200
				Total:	4,000

Each pairing from the 20 cases in Table 2 consists of a source image and a comparison image. The source image is the original image with no compression operations applied to it in that image's lifecycle. The comparison image from each of the 20 cases described in Table 2 was processed by compressing that image into each of the 14 ratios described in Table 3, and then decompressing the image. Compression is "lossy" in all cases except that for the 1:1 ratio (control case) where no compression was applied. This processed image has now passed through the compression and decompression stages of the algorithm representing what would be the typical lifecycle of the image in a normal operational environment just prior to processing/enrollment. Each image pair for every case in Table 2 is then presented to the examiners as described in section 2.2. Further details on the selection of images for this study are described in Appendix B.

It should also be noted that images may be decompressed many subsequent times, but the decompression process has no effect on the compressed stream and therefore not a topic of study. The images were compressed only once, thereby limiting compression degradation to only that first compression pass. Additional information on the makeup of the experimental image deck is provided in Appendix B.

The compressed images resulting from processing were then paired with the original image, creating a set of 56,000 image pairs. Thus one image of each pair is a copy of the original image, and the second is a copy of the processed image at one of the various compression ratios, including a control image as shown in Table 3.

Table 3 - Pairing of Each Original Images to Compressed Counterpart

Pair Number	Image 1 from pair	Image 2 from pair
1	Original / Non-compressed	1 to 1 (Non-compressed **control pair**)
2	Original / Non-compressed	2 to 1
3	Original / Non-compressed	5 to 1
4	Original / Non-compressed	7 to 1
5	Original / Non-compressed	10 to 1
6	Original / Non-compressed	12 to 1
7	Original / Non-compressed	15 to 1
8	Original / Non-compressed	17 to 1
9	Original / Non-compressed	20 to 1
10	Original / Non-compressed	22 to 1
11	Original / Non-compressed	26 to 1
12	Original / Non-compressed	30 to 1
13	Original / Non-compressed	34 to 1
14	Original / Non-compressed	38 to 1

2.1. Compression Algorithm

The focus of this study is to measure the effects of compression on fingerprint image fidelity from using the JPEG 2000 algorithm [JPEG2K] on fingerprint imagery. JPEG 2000 is an image compression standard and coding system that was created by the Joint Photographic Experts Group committee (JPEG) in 2000 to improve on the original JPEG image compression standard's discrete cosine transform-based methodology [JPEG] by utilizing a wavelet-based methodology. This modification yielded increases in both data compression and subjective image quality. Moreover, JPEG 2000 provides additional flexibility in the creation and manipulation of the code-stream and is based on the same family of wavelets as WSQ which is currently the standard for fingerprint image compression at 500 ppi. The flexibility offered by JPEG 2000 as well as the greater availability of JPEG 2000 implementations, which are commodity products as opposed to the much more specialized WSQ implementations, make JPEG 2000 a good candidate as the successor to WSQ.

The implementation of JPEG 2000 used in this experiment was Open JPEG's [OPENJPEG] reference implementation version 1.3. This reference implementation has been incorporated into the NIST Biometric Image Software (NBIS) public domain software distribution [NIST2].

For this experiment it was necessary to create a specially tailored compression approach to generate test images for all necessary compression ratios being examined. This tailoring strategy utilized several exploratory studies which established bounds and baselines for parameters such as the compression ratio, intermediate compression layers and decomposition levels. This tailoring yielded 13 sets of parameters, one set for each of the 13 compression ratios to be investigated. Each of the original images used was processed at each of the 13 compression levels, yielding 13 compressed images plus one control image that was not compressed. Thus each original image used in the study yielded 14 test images. The specially tailored compression approach was based on the guidance found in the informative "Profile for 1000 ppi Fingerprint Compression" [MTR1]. The guidance, which called for 6 decomposition levels, was adjusted to use 5 decomposition levels using data from the exploratory studies in creating the tailored parameter sets. These target compression ratios and associated intermediate compression levels are listed in Table 4. Note that JPEG 2000 enables structuring of the code stream such that it may be progressively decompressed at any of a series of intermediate compression levels in addition to the final "target" compression level. This feature of JPEG 2000 is intended to allow for the display of lower fidelity versions of the image to suit, for example, lower resolution displays while adding negligibly to the size of the compressed data stream and having no effect on the image at the target compression level. Other specific configuration parameters used for the compression of images are provided in Table 5.

Table 4 - Compression Levels Used in Study

Target Compression Ratio	Target and Intermediate Compression Levels[3]
1 to 1	Control image, no processing was performed
2 to 1	214, 144, 86, 58, 34, 24, 15, 10, 2
5 to 1	214, 144, 86, 58, 34, 24, 15, 10, 5
7 to 1	214, 144, 86, 58, 34, 24, 15, 10, 7
10 to 1	324, 214, 144, 86, 58, 34, 24, 15, 10
12 to 1	324, 214, 144, 86, 58, 34, 24, 15, 12
15 to 1	540, 324, 214, 144, 86, 58, 34, 24, 15
17 to 1	540, 324, 214, 144, 86, 58, 34, 24, 17
20 to 1	540, 324, 214, 144, 86, 58, 34, 24, 20
22 to 1	540, 324, 214, 144, 86, 58, 34, 24, 22
26 to 1	980, 540, 324, 214, 144, 86, 58, 34, 26
30 to 1	980, 540, 324, 214, 144, 86, 58, 34, 30
34 to 1	1930, 980, 540, 324, 214, 144, 86, 58, 34
38 to 1	1930, 980, 540, 324, 214, 144, 86, 58, 38

Table 5 - Other JPEG 2000 Compressor Settings Used in Study

Compressor Configuration Setting	Description
-n 6	6 resolution levels (original + 5 levels of decomposition[4])
-p RPCL	Resolution-Position-Component-Layer (RPCL) progression order
-b 64, 64	Code block size of 64x64
-r [rate values from Table 4]	Specifies the target top-layer rate, plus other quality layers
-d 0,0	Image origin offset
-I	Use irreversible compression (lossy)
-S 1,1	Use subsampling factor of 1,1

[3] Note that at higher target compression ratios, intermediate compression levels indicated in the guidance were found to cause malfunction of the Open JPEG 2000 v.1.3 encoder. Partially influenced by the image size, intermediate layers exceeding 324:1 to 540:1 caused the algorithm to return a minimally compressed image rather than the expected ratio. Accordingly, problem images were re-run using more modest intermediate layer specifications. The intermediate levels were found to have no effect on the output image at the target compression ratio. This issue has been corrected in v.1.4 of the Open JPEG 2000 encoder.

[4] The Open JPEG 2000 codec sets the number of decomposition levels to one less than the value specified by this command line parameter. Hence, -n 6 yields 5 decomposition levels.

2.2. Methodology

Each image pair from Table 2 was shown to exactly three examiners. Each examiner was first asked to determine if the image pair being displayed constitutes a matched pair from the same individual (referred to throughout this document as the "Identification decision" or "Non-Identification decision"). Their responses in determining the identity of the pair being presented to them can be one of three choices:

- The presented image pair is from the same individual ([positive] Identification decision);
- The presented image pair does not appear to be from the same individual (Non-Identification decision);
- Determination of identity cannot be made (Inconclusive).

Subsequent to their identity determination for the pair of images each examiner then evaluated the image pair on fidelity loss. To aid in analysis and quantification of fidelity loss, the examiner's evaluation was collected by utilizing a Likert-type response scale [LIKERT]. The choices that the examiners were allowed to make are provided in Table 6 below. Observation codes are ordered in ascending order indicating a progressively greater amount of degradation from 1 to 4. Furthermore, the features summarized in observation code 3 are among those typically used for forensic-level decisions while features summarized in observation code 4 are those used primarily by automated-matchers in rendering a match decision.

Table 6 - Observation Codes for Compression Degradation Observation

Observation Numeric Code	Description
1	No apparent image quality degradation and the quality of Level II (2) [5] and Level III (3) detail in either image should not cause any difficulty in reaching a conclusive decision of identification or exclusion.
2	A noticeable degradation in the quality of Level II (2) or Level III (3) detail in either image, but not enough to have a negative impact on reaching a conclusive decision of identification or exclusion, though the amount of time to reach a decision may increase.
3	Level III (3) detail quality diminished in either image to the extent that a Level III (3) identification is questionable or not possible, and/or is significantly more difficult.
4	Level II (2) detail quality diminished in either image to the extent that a Level II (2) identification becomes questionable or not possible, and/or is significantly more difficult.

Examiner responses were recorded by custom test apparatus consisting of a commodity computer and software designed and developed specifically for this study. The examiners were not provided any time limits on their responses.

The 56000 image pairs were queued on each examiner's workstation and their presentation order was shuffled randomly on each of the three workstations.

The examiners were provided the basic ability to independently reposition, rotate, invert and zoom in and out of each of the two images from the pair being examined. This provided them with the basic tools that they typically employ in their standard operating environments in performing their duties. While they were provided with basic tools, more advanced assistive technologies normally available to some examiners, such as on-screen feature marking or image adjustments were not provided to them in the interest of experimental control.

---

[5] The commonly accepted nomenclature defines Level 1 fingerprint details as the overall friction ridge pattern and flow, Level 2 detail as classic Galton features [GALTON] like minutiae points, and Level 3 as pores, creases, line shapes, incipient ridges and other non Level 1 or 2 features [JAIN].

The image pairs were presented on split-screen to the examiners in randomized order (see Figure 3), and randomized placement (left/right split screen placement) to mitigate potential order effects or positional bias. Scientists overseeing the tests were blind to the placement order of the images, as well as to the compression level of image pairs. These factors were tracked by the test apparatus.

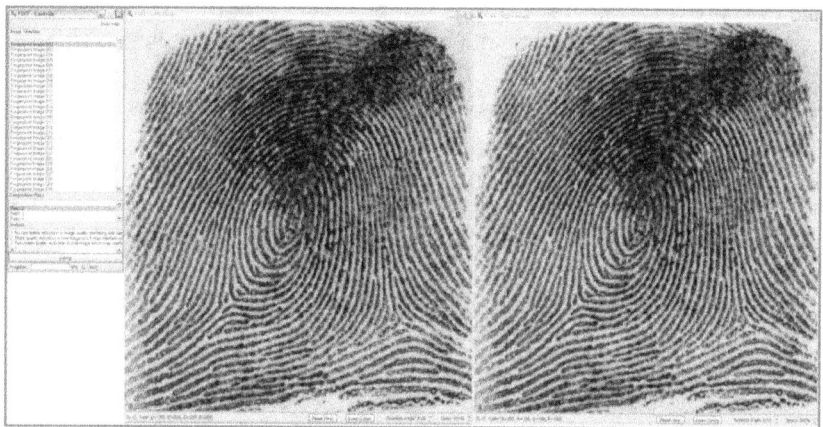

Figure 3 - Split-Screen Presentation of Image Pairs

Each of the 56000 image pairs was guaranteed to be observed by 3 different examiners over the course of the study without repeating. Due to the physical limitations of even the best modern monitors, it is impossible to show a 1000 ppi image without zooming or interpolation. The software apparatus enabled the examiner to view images at approximately 10x to 50x of the original size (see Appendix C for more information).

Once the examiner made a determination for a given pair of images, that pair was marked clearly as complete. The examiner was allowed to return to a completed pair and re-examine that image pair without penalty. The examiner was also allowed to jump to any image pair in the queue regardless of that pair's position in the examiner's queue.

Examiners were provided basic verbal instructions and a demonstration on how to use their workstations and were allowed a brief practice session using image pairings that were not part of the study to gain familiarity with the procedure, scoring, and workstation controls. Examiners were allowed to freely ask questions or clarification on their workstations or tasking. The examiners were located in the same room and were allowed to interact freely as they do in their normal professional practice. Finally, the examiners were advised that once they've selected one of the three workstations on which to process images, they continue to use that workstation exclusively. This was done in order to eliminate the possibility of an examiner processing an image pair more than once.

2.3. Participants

This study utilized 39 paid professional fingerprint examiners to look at fingerprint image pairs in their professional capacity and to render their professional judgment much as they do in their normal professional activities.

There was no attempt to evaluate the examiner's accuracy or proficiency for this study either prior, during or after the study. There was also no attempt to identify or maintain the identity of the individual examiners utilized in this study.

The examiners had anywhere from two years to well over forty years of experience in fingerprint examination.

A pre-requisite for examiners being selected to participate in this study was that they be trained *latent-print-examiners (LPE's)*. This requirement ensures that the examiners have received the most rigorous training possible and can resolve even the most difficult fingerprint identification cases which are typically from latent fingerprint images that, unlike controlled fingerprint captures, are typically fragmentary and often distorted. Of the 39 examiners participating in this study, 90 % had earned the IAI's latent examiner certification (and were referred to as *certified latent print examiners*, or *CLPE's*) with the remaining four being non-certified latent examiners. Selection of examiners for this study was not meant to bias the results by utilizing latent-only examiners and all of the examiners selected for this study perform 10-print case work in addition to latent case work as part of their regular professional duties. While not the case in this study, it should be noted that some of LPE's or CLPE's working with larger agencies may be assigned latent-only case work with almost no 10-print case work.

The examiners were recruited from various federal, state, local and commercial entities and were permanently based in 16 states: Alabama, Arkansas, California, Colorado, Florida, Illinois, Iowa, Louisiana, Maryland, Mississippi, Missouri, North Carolina, Pennsylvania, South Carolina, Texas, and Virginia.

3. Analysis

3.1. Observed Image Quality

After the completion of data collection a certain degree of preprocessing had to be performed on the raw data in order to aid in the analysis of the results. In the case of compression observations, the observation selections by the examiners were captured by the test apparatus as numeric codes as indicated in Table 6 above.

Expectation based on anecdotal evidence suggested that with increasing compression one should observe a general trend in ratings from "1" toward ratings of "4" indicating increasing amounts of observed image degradation due to compression. Indeed this is observed to be the case as demonstrated by Figure 4 below.

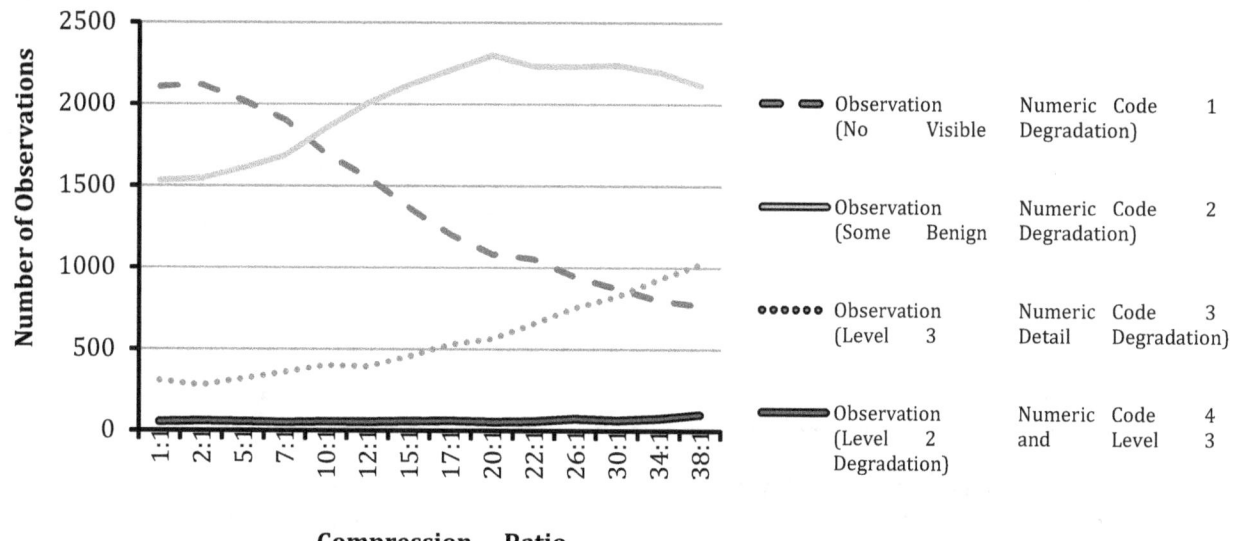

Figure 4 - Trending of Responses with Increasing Compression

This trending pattern is generally true for every fingerprint pairing scenario (Table 2) where the frequency of higher rating codes increases with increasing compression rate. What is not clear in this is the statistical significance of the trends observed in image degradation with increasing compression.

Each rating consists of a composite of response codes by the three examiners. Several methods were tested to determine the best means by which to represent the 3-valued composite evaluations. In the case of unanimous assignment of response codes by all examiners, there is no ambiguity in their observation. However, investigators had to determine the best way to reconcile any disagreement among examiners. That is, how can the 3-valued ratings be ordered to reflect qualitative and quantitative levels of image degradation? Given 20 possible 3-way ratings from "1,1,1" to "4,4,4" as shown in Table 7, one method explored was to assign an average cost, a real value ranging from 0 to 1, to each of the 20 possible ratings combinations.

Based on further examination of how the IAI study utilized a majority-rule decision in their study [FITZPATRICK], a "majority-rule" partitioning of cost was devised as shown in Table 7 where the 20 cases were collapsed into 4 cases. The legacy IAI study utilized a mechanism similar to this, but they utilized 2 examiner decision points and employed a third examiner/decision only where a deadlock was reached, where the third examiner was used to establish consensus and break the deadlock.

Table 7 - Normalized Ratings and Assigned Costs

Responses from Observers	Normalized Rating	Associated Cost	Justification
1,1,1 1,1,2 1,1,3 1,1,4	0	0	These cases represent either a unanimous or majority ruling of observation code 1 from Table 6 indicating no apparent image degradation.
1,2,2 1,2,3 1,2,4 2,2,2 2,2,3 2,2,4	1	0.25	These cases represent either a unanimous or majority ruling of observation code 2 from Table 6. This case also includes the special split-decision cases of 1, 2, 3 and 1, 2, 4. These cases are considered borderline cases but are assigned to this bin as they are biased towards an acceptable image rating by two examiners noting little or no observable loss.
1,3,4 2,3,4 1,3,3 2,3,3 3,3,3 3,3,4	2	0.5	These cases represent either a unanimous or majority ruling of observation code 3 from Table 6. This case also includes the special split-decision cases of 1, 3, 4 and 2, 3, 4. These cases are considered biased-towards, and indicative of level-3 detail loss.
1,4,4 2,4,4 3,4,4 4,4,4	3	1	These cases represent either a unanimous or majority ruling of observation code 4 from Table 6.

For every condition of each fingerprint type and compression ratio, we have 200 3-value ratings. Each 3-value rating is converted to a single normalized rating (Table 7) reflective of the degree to which one of the fingerprints is judged to have lost some features useful for identification. From these 200 ratings, we count the frequency, f_i, of each of the normalized rating values ($i = 0...3$) and convert each frequency to a probability by dividing frequency by 200 (equation (1)).

$$\frac{f_i}{200} \quad i = 0...3 \tag{1}$$

$$\tag{2}$$

Then using the cost values specified in Table 7 as a measure of the magnitude of degradation we compute a degradation score (Q) observed for a particular condition as the sum of the products of probabilities of ratings and cost values assigned to the ratings as shown in equation (2).

Thus, for each matching scenario and compression level, the value of the score, Q, provides a summary measure of the degree of degradation observed among the 200 image pairs for the condition under examination. As the 200 measurements for each matching scenario and compression level combination yield only a single value of the degradation score (Q), a bootstrap procedure as described in section 3.1.3 was used to generate a distribution of such values (bootstrap replicates) from which to estimate uncertainties and parameters to be used in hypothesis tests of differences among experimental conditions.

3.1.1. Normality of Degradation Score

In order to apply hypothesis tests as described in the next section, we examined the degree to which the distributions of bootstrap replicates (see section 3.1.3) of the degradation score approximate a standard normal distribution. Attempts to apply quantitative tests of normality such as the Shapiro-Wilks normality test [SHAPIRO] proved problematic due to the quantized nature of the scores. Such tests compare a smooth, continuous cumulative normal distribution to similarly behaved data under test. Our data originate from frequency counts of categorical examiner responses weighted by a limited number of cost values. The resulting distributions, though normal in shape, have cumulative distribution functions composed of a series of discrete steps to which quantitative tests of normality are quite sensitive. Accordingly, we applied and accept the normality assumption on the basis of inspection of histograms with overlain normal fitted functions and normal probability plots. Typical examples are displayed in Figures 5 and 6 below. The normal probability plot [CHAMBERS] provides a nonparametric means by which to compare quantiles of two distributions. Distributions may be taken as normal if a quantile plot overlays that of the standard normal distribution (a straight line in this type of plot) without major departures from linearity.

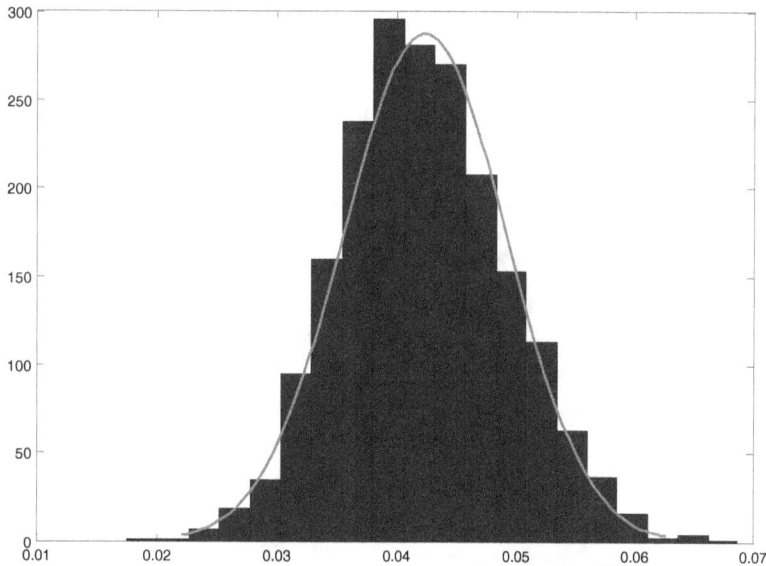

Figure 5 - Histogram typical of degradation score data replicates output from bootstrap procedure with the normal distribution function fitted to data.

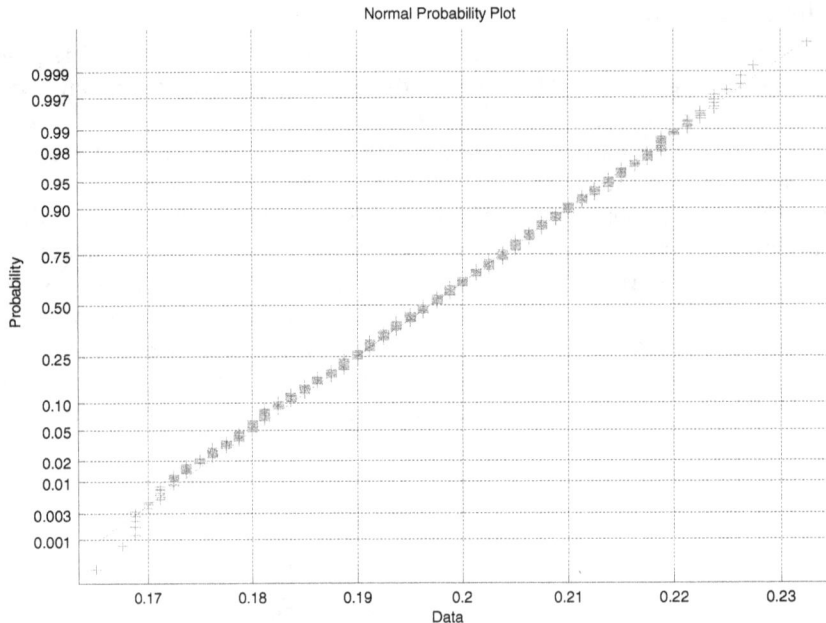

Figure 6 - Normal probability plot typical of degradation score data. Normality is indicated by linearity.

3.1.2. Hypothesis Testing

It is important to keep in mind that all values of the degradation scores reflect the relative observed degradation of a *comparison* image with respect to an original source image. In all cases the source image is not compressed. The comparison image may or may not have undergone compression processing, and may be a match or non-match fingerprint.

Specifically, the comparison image may be a compressed version of the source image (same person/same impression), a compressed version of a match image (same person, same finger, but impression taken at different time), or a compressed image of a non-match (impostor) fingerprint (different person, any finger pairing). For experimental control, every source image in all scenarios includes a comparison of the non-compressed source image with itself.

Thus, every comparison includes a non-compressed source image compared with another image at each of the 14 compression levels from Table 4 including the non-compressed/control image, also referred to as the 1:1 ratio. Accordingly, the data supports hypothesis testing of the degradation contrast between non-compressed images and paired comparison images having compression ratios from 1:1 to 38:1 as described previously. Contrast between non-compressed, identical, image pairs serves as a control and as the baseline for evaluation of degradation observed at other compression ratios, as well as an important indicator of the inherent quality of that particular fingerprint impression.

Hence, the control data enables testing the hypothesis that degradation at each of the compression levels is equal to that observed among comparisons of non-compressed images. Inasmuch as each set of 200 rating scores from Table 2 yields only a single value of the degradation score, Q, we use a bootstrap procedure as described in the next section to estimate the values of the test statistic (the mean of Q for bootstrap replicates) and uncertainty (the standard error for bootstrap replicates) to be used for hypothesis testing. The difference between the values of the test statistic for any two compression levels may be tested in the manner applied to comparing different algorithms as described by Wu, et al. [WU1].

For each of the 20 matching scenarios from Table 2, $m = 1\ldots20$, we let $Q_{1,m}$ and $Q_{n,m}$ denote the score, Q, as defined above for the control case (1:1) and compression case respectively, where $n = 2\ldots14$ compression ratios from Table 4. The null and alternative hypotheses, then, are

$$H_{01,nm}^{nm} : Q_{1,m} = Q_{n,m}$$
$$H_{a,nm}^{nm} : Q_{1,m} \neq Q_{n,m}$$

Assuming normality of these statistics, the Z test statistic is

$$Z_{nm} = \frac{\hat{Q}_{1,m} - \hat{Q}_{n,m}}{\sqrt{SE^2(\hat{Q}_{1,m}) + SE^2(\hat{Q}_{n,m}) - 2 r_{1,nmm} SE_{1,m} SE_{n,m}}} \quad (3)$$

where $\hat{Q}_{1,m}$ and $\hat{Q}_{n,m}$ are estimators (means) of the scores, $SE\hat{Q}_{1,m}$ and $SE\hat{Q}_{n,m}$ are the standard errors of the two estimators of the Q scores, and $r_{1,nmm}$ is the correlation coefficient between $Q_{1,mk}$ and $Q_{n,mk}$, where $k = 1\ldots K$ bootstrap replicates.

3.1.3. Bootstrap Procedure

Given 200 observations (normalized ratings) for each of the 20 matching scenarios (see Table 2) at each of the 14 compression levels (Table 4), the values of the test statistic, $\hat{Q}_{1,m}$ and $\hat{Q}_{n,m}$, the standard errors, $SE\hat{Q}_{1,m}$ and $SE\hat{Q}_{n,m}$, and corresponding $r_{1,nmm}$ values are computed using a non-parametric two-sample bootstrap procedure [WU2].

The bootstrap procedure consists of K=2000 executions of the procedure involving equations (1) and (2) above, each run using a uniform random sample of 200 cases drawn, with replacement, from the original 200 observations. The result of the bootstrap procedure is a matrix of score values, Q_{kn}, k=1...2000 bootstrap replicates and n=1...14 for the 14 compression levels. Such a matrix was computed for each of the 20 fingerprint type matching conditions from Table 2. For each fingerprint type and compression level, statistics are computed including the mean score value and standard error of the mean. Moreover, for each scenario the correlation is computed between the first column of the matrix of values and each of the other columns. Following the procedure described above, Z-scores were computed and are shown in Table 8. These results together with descriptive statistics supporting this analysis appear in Appendix E.

Table 8 - Z-Scores

		1:1	2:1	5:1	7:1	10:1	12:1	15:1	17:1	20:1	22:1	26:1	30:1	34:1	38:1
Ink Card Scan	Rolled to Rolled, Match Pair	---	0.44	-2.07	-4.56	-7.52	-9.46	-9.61	-11.94	-12.93	-15.38	-14.85	-15.92	-18.60	-17.02
	Rolled to Rolled, Non-Match Pair	---	0.16	-0.40	-1.12	-2.97	-1.85	-3.57	-4.46	-5.36	-6.96	-7.34	-6.63	-8.93	-8.43
	Rolled to Flat, Match Pair	---	0.80	-0.07	-1.59	-3.11	-2.63	-3.75	-4.96	-4.48	-5.19	-6.06	-6.74	-7.58	-8.46
	Rolled to Flat, Non-Match Pair	---	-0.62	-1.23	-1.29	-2.22	-1.79	-3.50	-4.07	-3.16	-5.15	-6.04	-6.61	-8.79	-10.60
	Flat to Rolled, Match Pair	---	-2.14	-1.69	-1.59	-3.27	-2.39	-3.25	-4.12	-5.36	-5.60	-5.58	-6.95	-8.44	-9.39
	Flat to Rolled, Non-Match Pair	---	-1.68	-1.89	-2.94	-3.17	-2.80	-6.02	-4.86	-5.18	-5.80	-5.94	-7.59	-8.33	-8.42
	Flat to Flat, Match Pair	---	0.75	-2.53	-0.79	-2.91	-5.32	-6.28	-7.91	-8.36	-11.56	-12.04	-10.33	-12.58	-13.34
	Flat to Flat, Non-Match Pair	---	1.39	-0.98	-1.04	-0.85	-2.07	-2.05	-3.86	-4.61	-4.48	-4.26	-5.51	-9.36	-8.28
	Slap to Slap, Match Pair	---	-0.02	-0.32	-1.35	-2.70	-5.13	-6.28	-7.54	-10.10	-9.56	-13.34	-13.37	-13.20	-13.78
	Slap to Slap, Non-Match Pair	---	0.17	-0.58	-1.11	-1.00	-0.21	-0.57	-0.39	-2.42	-0.73	-4.24	-2.95	-2.96	-5.25
Digital Live Capture	Rolled to Rolled, Match Pair	---	0.98	-0.11	-1.74	-3.45	-7.60	-8.34	-11.99	-12.58	-15.29	-16.56	-17.22	-17.90	-18.01
	Rolled to Rolled, Non-Match Pair	---	-0.35	-0.87	-1.89	-3.97	-4.91	-5.75	-6.41	-8.50	-8.69	-9.25	-9.02	-10.16	-10.30
	Rolled to Flat, Match Pair	---	0.97	-0.12	1.26	0.37	0.13	-0.85	-1.50	-0.43	-0.95	-1.06	-2.14	-1.46	-1.81
	Rolled to Flat, Non-Match Pair	---	0.96	-0.25	-0.02	0.84	0.15	-0.41	-0.54	-1.37	-1.79	-1.23	-1.97	-1.91	-1.98
	Flat to Rolled, Match Pair	---	-0.17	-1.23	-0.79	-1.66	-0.35	-1.36	-1.85	-1.24	-1.43	-2.08	-2.84	-0.67	-2.24
	Flat to Rolled, Non-Match Pair	---	1.30	-0.43	0.06	-0.89	-0.48	0.98	-0.77	-0.64	0.43	-1.23	0.32	-1.00	-2.39
	Flat to Flat, Match Pair	---	-0.37	-1.04	-1.18	-2.31	-2.33	-3.89	-5.40	-5.89	-4.69	-6.70	-6.99	-7.55	-6.32
	Flat to Flat, Non-Match Pair	---	-0.34	-0.63	-0.48	-1.30	-2.93	-2.13	-3.17	-2.87	-3.28	-4.60	-4.78	-5.41	-5.12
	Slap to Slap, Match Pair	---	1.03	-0.66	-0.98	-3.67	-6.77	-10.19	-12.89	-16.80	-17.16	-19.49	-20.24	-21.20	-23.58
	Slap to Slap, Non-Match Pair	---	0.37	-0.50	-0.23	-3.41	-4.71	-9.28	-9.76	-12.10	-11.89	-12.96	-14.12	-16.85	-16.90

The Z-scores correspond to standard deviation units of the standard normal distribution which has a mean of zero and a standard deviation of one. As one may recall from basic statistics, standard deviation units mark off areas of the standard normal distribution that correspond to probabilities. Hence, 95 % of the distribution lies between Z-scores ±1.96. In our two-sample Z-test, we apply the Z-test to assess the difference between $\hat{Q}_{1,m}$ and $\hat{Q}_{nm}$. A Z-score, computed using equation (3), greater than +1.96 or less than -1.96 would indicate less than a 5 % chance (p<0.05) that two samples drawn from the same population could differ by the observed amount. We have set the probability threshold, or α level, beyond which we will reject the null hypothesis at α=0.05. That is, we reject the null hypothesis if p<0.05 and accept the alternative hypothesis that the two values of the comparison statistic are significantly different and, by implication, that the algorithms under examination, i.e. the two compression rates, produce different results.

Table 9 shows probability values corresponding to the Z-scores of Table 8. Note that inasmuch as we allow that the difference between the two statistic estimators may be either positive or negative, we distribute the 5 % rejection region between the two tails of the distribution, hence the hypothesis test is a "two-tailed" test.

3.1.4. Two Tailed Test

Table 9 displays the results of comparisons of the test statistic (p-values) among all match scenarios and compression levels for the ink card scan images and digital live capture fingerprint images. As indicated above, each of the comparisons tests significance of the contrast between the degradation score Q for the control case of pairs of non-compressed images and those for which one of the images has been compressed at one of the ratios exclusive of the 1:1. Thus, probability values less than the $\alpha=0.05$ level indicate significant differences between perceived image degradation relative to that for non -compressed images. Inspection of Table 9 indicates that for most of the scenarios, the significant degradation beyond that of non-compressed imagery first occurs at 10:1 compression.

Table 9 - P-Values for 2-Tailed Tests

		1:1	2:1	5:1	7:1	10:1	12:1	15:1	17:1	20:1	22:1	26:1	30:1	34:1	38:1
Ink Card Scan	Rolled to Rolled, Match Pair	---	0.657	0.039	0.000	0.000	0.000	0.000	0.000	0.000	0.000	0.000	0.000	0.000	0.000
	Rolled to Rolled, Non-Match Pair	---	0.869	0.689	0.262	0.003	0.064	0.000	0.000	0.000	0.000	0.000	0.000	0.000	0.000
	Rolled to Flat, Match Pair	---	0.424	0.942	0.113	0.002	0.009	0.000	0.000	0.000	0.000	0.000	0.000	0.000	0.000
	Rolled to Flat, Non-Match Pair	---	0.534	0.218	0.197	0.027	0.073	0.000	0.000	0.002	0.000	0.000	0.000	0.000	0.000
	Flat to Rolled, Match Pair	---	0.032	0.092	0.111	0.001	0.017	0.001	0.000	0.000	0.000	0.000	0.000	0.000	0.000
	Flat to Rolled, Non-Match Pair	---	0.092	0.059	0.003	0.001	0.005	0.000	0.000	0.000	0.000	0.000	0.000	0.000	0.000
	Flat to Flat, Match Pair	---	0.451	0.011	0.427	0.004	0.000	0.000	0.000	0.000	0.000	0.000	0.000	0.000	0.000
	Flat to Flat, Non-Match Pair	---	0.163	0.325	0.297	0.396	0.038	0.041	0.000	0.000	0.000	0.000	0.000	0.000	0.000
	Slap to Slap, Match Pair	---	0.980	0.748	0.177	0.002	0.000	0.000	0.000	0.000	0.000	0.000	0.000	0.000	0.000
	Slap to Slap, Non-Match Pair	---	0.862	0.560	0.266	0.317	0.837	0.567	0.698	0.016	0.467	0.000	0.003	0.003	0.000
Digital Live Capture	Rolled to Rolled, Match Pair	---	0.326	0.915	0.082	0.001	0.000	0.000	0.000	0.000	0.000	0.000	0.000	0.000	0.000
	Rolled to Rolled, Non-Match Pair	---	0.725	0.383	0.059	0.000	0.000	0.000	0.000	0.000	0.000	0.000	0.000	0.000	0.000
	Rolled to Flat, Match Pair	---	0.330	0.903	0.206	0.713	0.895	0.393	0.134	0.666	0.343	0.291	0.032	0.145	0.070
	Rolled to Flat, Non-Match Pair	---	0.336	0.802	0.984	0.403	0.879	0.682	0.588	0.170	0.074	0.217	0.049	0.056	0.047
	Flat to Rolled, Match Pair	---	0.864	0.218	0.430	0.098	0.724	0.174	0.064	0.214	0.154	0.037	0.004	0.504	0.025
	Flat to Rolled, Non-Match Pair	---	0.192	0.666	0.956	0.372	0.630	0.328	0.444	0.523	0.667	0.218	0.750	0.315	0.017
	Flat to Flat, Match Pair	---	0.710	0.298	0.240	0.021	0.020	0.000	0.000	0.000	0.000	0.000	0.000	0.000	0.000
	Flat to Flat, Non-Match Pair	---	0.738	0.530	0.628	0.195	0.003	0.034	0.002	0.004	0.001	0.000	0.000	0.000	0.000
	Slap to Slap, Match Pair	---	0.304	0.510	0.327	0.000	0.000	0.000	0.000	0.000	0.000	0.000	0.000	0.000	0.000
	Slap to Slap, Non-Match Pair	---	0.714	0.616	0.815	0.001	0.000	0.000	0.000	0.000	0.000	0.000	0.000	0.000	0.000

Earlier in this section, Figure 4 provided anecdotal evidence of a monotonically increasing number of observed compression degradation as the level of compression increases. The Z-scores presented in Table 8 reinforces this progressive increase of observed compression anomalies as the greater compression ratios are utilized on images. Yet, it has been suggested that for inked prints, a small degree of compression can improve image quality by filtering the image of some noise. Several plots such as Figure 7 below (inked roll-to-roll match case) appear to reinforce this notion, e.g. by virtue of the slight dip in perceived degradation from 1:1 to 2:1 compression ratios. However, these data generally do not provide evidence of a statistically significant improvement in image quality at very low compression ratios (such as 2:1) over the non-compressed image. (The exception is the inked, flat-to-flat, match pair case that shows a positive z-score at 2:1, increasing to become significant at 5:1 with $p<0.05$.)

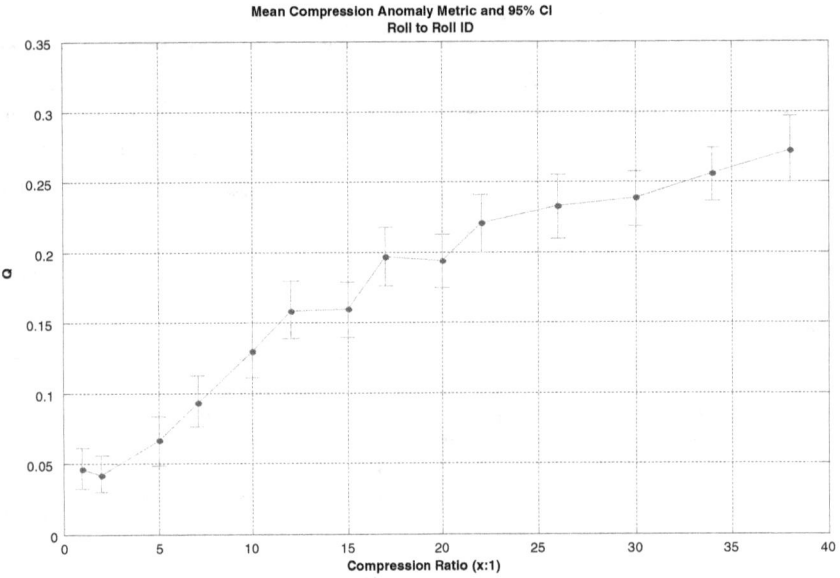

Figure 7 - Mean compression anomaly measure with 95 % confidence intervals for inked, rolled-to-rolled, fingerprints (match pairs) for compression ratios from 1:1 to 38:1.

3.2. Identification Errors

In addition to studying the effects of compression and the resulting image anomalies, this experiment also studied the effects of compression on the ability of the examiner to make an identification decision. Investigators expected that even at the most extreme level of JPEG 2000 compression used in the present study, the rate of misidentification would be low among the experienced examiners. However, the task was included in the study to capture possible increase in some nominal misclassification rate with increased compression.

As with the analysis on observed compression anomalies, normalization was performed on the individual decisions to allow for them to be analyzed as a singular unit for each image pair. Prior to normalizing decision triplets into a single decision value, the individual decisions were assigned a code. In the case of identity determination, the codes are provided in Table 10 below.

Table 10 - Observation Codes for Determination of Identity

Observation Code	Description
Y	A correct determination of identity was made.
N	A correct determination of non-identity was made.
U	A determination of identity was not made given the image pair.

For each pair of images examined, a match decision was provided by 3 different examiners. The decision triplets were then categorized into a singular code and each was assigned a normalized code. For example, decision "YYN", "NYY" and "YNY" were all treated as and a single decision of "YYN", and then assigned a normalized code of "0" (from Table 11) Categorization of identity determination triplets are provided in Table 11 below.

Table 11 - Normalization Table for Identification Decision

Responses from Observers	Normalized Code	Justification
Y,Y,Y Y,Y,N Y,Y,U	0	These cases represent a majority-identification decision by the examiners.
N,N,N N,N,Y N,N,U	1	These cases represent a majority-non-identification decision by the examiners.
N,U,Y U,U,Y N,U,U U,U,U	2	These cases represent a majority-inconclusive decision by the examiners. Split decision cases (N, U, Y) are also considered inconclusive in this study.

3.2.1. Analysis of Identification Error Rates

In order to examine the extent of misclassification, we take advantage of the fact that for each of the matching conditions, e.g. slap-to-slap, flat-to-roll, etc. we have equal numbers of identification and non-identification pairings. Thus, we are able to examine error rates in a conventional way, i.e. considering overall error to include cases in which known identification cases are classified as non-identification, false negatives (type II error), and the cases in which known non-identifications are classified as identifications, false positives (type I error). In order to deal with the classification as a binary decision, we must include the inconclusive decisions with one or the other category.

In conventional fingerprint examination, the conservative approach is to consider an inconclusive determination as a "correct" response. That is, in the absence of a high degree of certainty over the identification, the preferable approach is to consider an inconclusive determination to be the correct response. However, for purposes of the present analysis, we are also interested in any possible increase in the number of inconclusive determinations with increasing compression level. Accordingly, in the analysis to follow, we treat the inconclusive classifications as correct responses. Then, in a separate analysis we examine the frequency of inconclusive responses among identification and non-identification pairings over compression level.

Given frequency distributions of identification determinations for each of the 10 fingerprint matching cases [6], j, and 14 compression ratios, k, we can define the error rate as

$$E_{jk} = \frac{fpr_{jkjk} + fnr}{2} \tag{4}$$

where

$$fpr = \frac{falsepositives}{falsepositives + truenegatives} \tag{5}$$

and

$$fnr = \frac{falsenegatives}{falsenegatives + truepositives} \tag{6}$$

The error rates are determined by counting the numbers of erroneous classifications for the 200 match and non-match pairs of each fingerprint type. The confidence intervals for the error rates are estimated using two approaches described in the following.

Where the number of observed errors is less than or equal to 3 out of the total 200 trials for both match or non-match cases, we use a 95 % confidence interval derived from properties of the binomial distribution rather than from quantiles of the error estimate distribution. The rationale for and derivation of this approach is elaborated in references such as [EYPASCH, HANLEY, and JOVANOVIV]. The so-called "Rule of Three" derives from the notion that in experiments involving binomial trials, the observation of zero errors (or adverse effects) does not imply that the actual failure rate is zero. To accept such an estimate of the failure rate would violate standards of conservatism in such studies typical of medical research. Given zero errors (or failures) in n observations, we can be 95 % confident that the actual error rate will lie between 0 and $3/n$. In our case, this upper limit corresponds to $3/200 = 0.015$ or 1.5 %. For cases in which the average number of observed errors is less than or equal to three, but greater than zero, we extend the rule of three such that confidence intervals become $[0, 4/n]$, $[0, 6/n]$, and $[0, 7/n]$ for average [7] error frequencies of 1, 2, and 3, respectively, to augment the case $[0, 3/n]$ for observed error = 0, giving us four "Extended Rule of Three" confidence intervals.

For cases in which the error frequency of either false negatives (fn) or false positives (fp) are greater than 3, we use the error rate estimate computed according to equations (5) or (6) and apply a bootstrap procedure to estimate the limits of the confidence interval that contains this error estimate. The bootstrap procedure samples with replacement from the 200 match and/or non-match pair cases, tabulates the false negatives and/or false positives and computes a

---

[6] We have combined the identification and non-identification scenarios for the error analysis, hence 20 mentioned earlier reduces to 10.

[7] Here we average the frequencies of false negatives and false positives, $\frac{fnfp}{2}$, and round the result to the nearest integer.

bootstrap replicate of the error estimate. Each iteration of the bootstrap combines the observed error frequency found to be less than or equal to three with a new count of false negatives or false positives from the sample. The bootstrap procedure [WU$_2$] is applied only when number of errors exceeds 3 in 200 for false positives, false negatives, or both. For example, if $fn \leq 3$ and $fp > 3$, fnr computed via equation (4) is held constant while a new value of fpr is calculated for each iteration of the bootstrap, i.e., for each sample of non-match pair cases. A composite error replicate is computed according to equation (4). If $fn > 3$ and $fp \leq 3$, we hold the fp constant and apply the bootstrap sampling to the match pair cases. If $fn > 3$ and $fp > 3$ we apply the bootstrap to both. In all cases, we generate a distribution of 2000 replicates of the error estimate. Quantiles 0.025 and 0.975 of this distribution mark the lower and upper limits of the 95 % confidence interval, i.e. for a two-tailed significance test with $\alpha = 0.05$.

In spite of the rather elaborate analysis protocol developed to examine identification errors, we observed only a very small number of misidentifications. In the graphs below we plot for each fingerprint comparison modality the match/non-match pair error estimate with 95 % confidence intervals for each of the 14 compression ratios considered in the present study. In all cases error estimates were found to fall within the region covered by the "Extended Rule of Three" as described above. Error was either zero, with confidence interval [0, 3/n], or one, with confidence interval [0, 4/n]. The only scenarios for which any error is observed involved rolled-to-flat or flat-to-rolled comparisons as shown in Figures 8 – 10 below. In each case the error estimate is plotted with confidence limits for each of the compression ratios. No clear trend is seen over increasing compression, and the few errors include only false negatives, i.e., cases in which identification was misclassified as non-identification. Inspection of the fingerprint pairs involved in the misclassification generally revealed ambiguity over the region of meaningful overlap between the rolled print and the corresponding flat impression. Moreover, the image manipulation "toolkit" provided to the examiners did not facilitate overlays or include other more sophisticated tools that might assist in comparing fingerprint features.

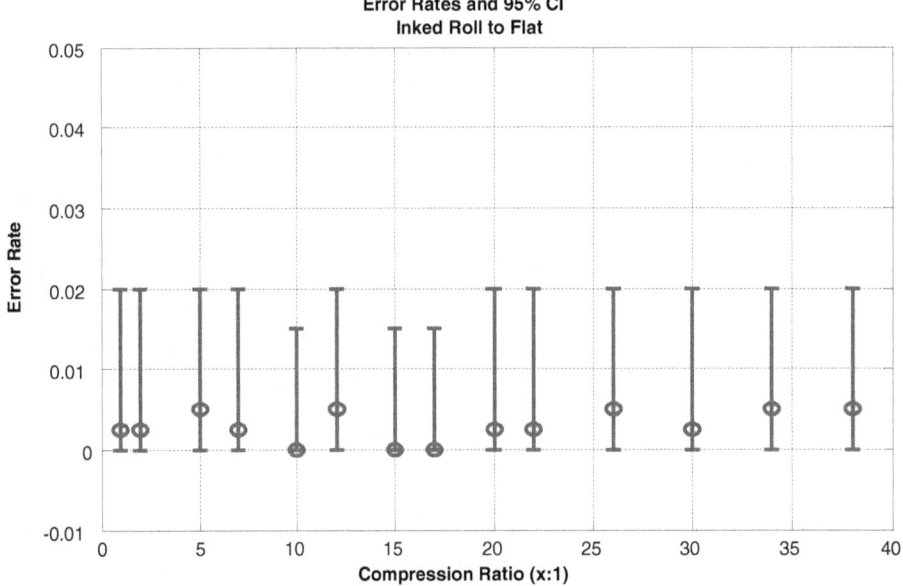

Figure 8 - Identification error rates for ink card scan rolled to ink card scan flat fingerprint comparisons at the 14 compression levels with confidence limits.

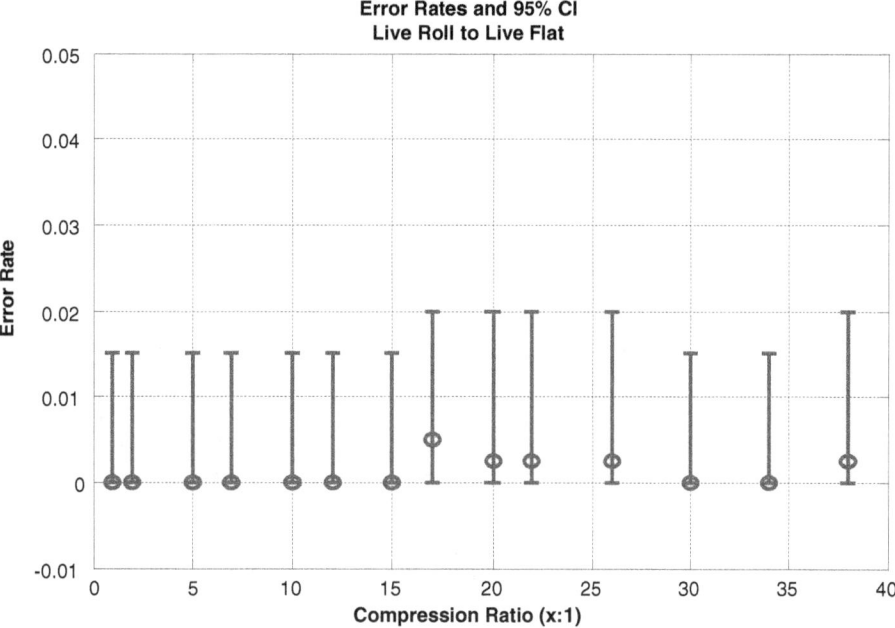

Figure 9 - Identification error rates for digital live capture rolled to digital live capture flat fingerprint comparisons at the 14 compression levels with confidence limits.

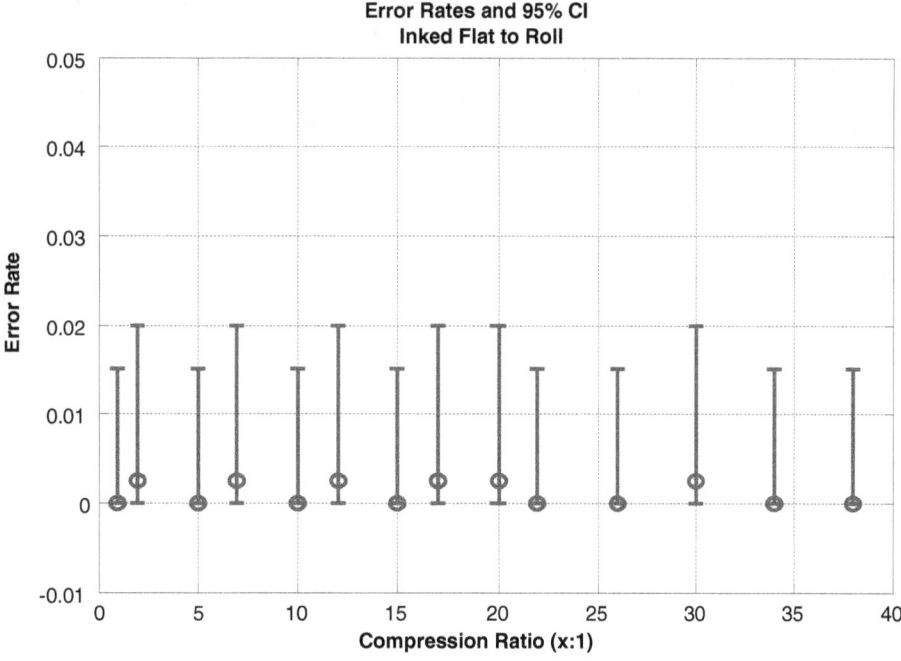

Figure 10 - Identification error rates for ink card scan flat to ink card scan rolled fingerprint comparisons at the 14 compression levels with confidence limits.

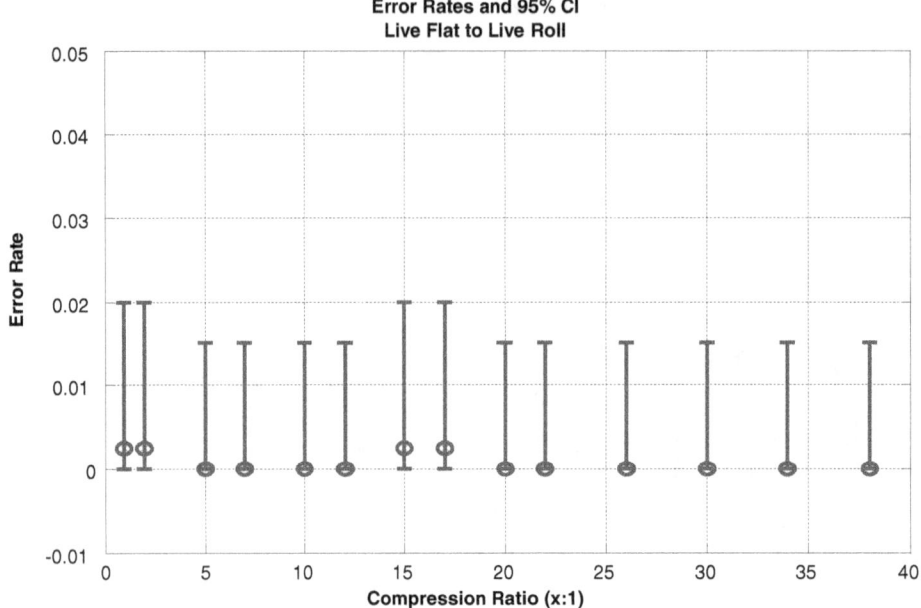

Figure 11 - Identification error rates for digital live capture flat to digital live capture rolled fingerprint comparisons at the 14 compression levels with confidence limits.

Thus, for all 10 fingerprint matching conditions and 14 compression levels from lossless to 38:1 we see mainly no error in making the identification/non-identification classification. Where error is observed, it does not exceed one comparison out of 200. In all cases, we have used the Rule of Three, and extensions thereof, to present the most conservative estimates of the 95 % confidence intervals for low probability events. Also, examination of Figure 8 through Figure 11 provides some anecdotal evidence that the identification error rates for ink card scan images may exhibit more variance than the digital live scan images at lower compression ratios.

3.2.2. Analysis of Inconclusive Cases

As to the possibility that compression might inhibit the ability of the examiners to make an identification decision, we examine the frequency of inconclusive determinations over compression levels for each of the fingerprint comparison modes. For this analysis we compare for each of the cases in Table 2 the relative probabilities of an inconclusive determination over the 14 compression levels from Table 4. In Figure 12 below, we observe that while there is some variability in the probability of an inconclusive decision on the part of examiners, particularly for the inked prints and comparison of rolled to flat impressions, there is no clear trend toward an increase in the probability of an inconclusive determination with increasing compression level.

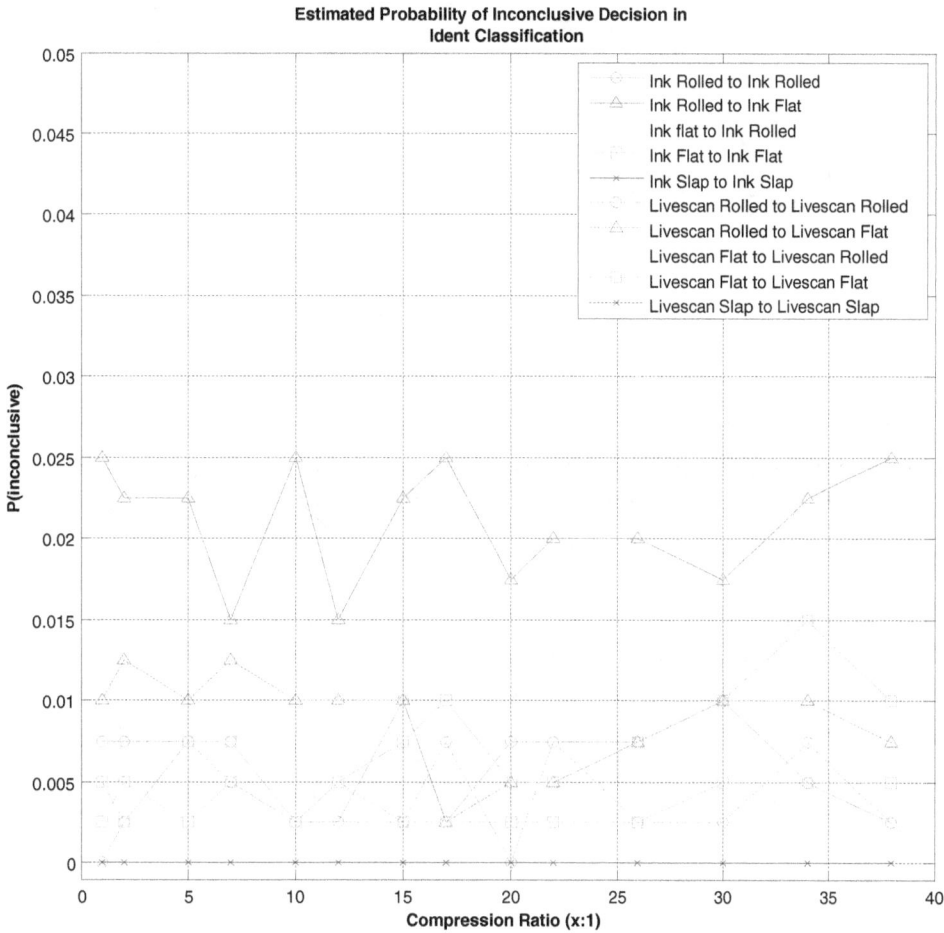

Figure 12 - **Probability of inconclusive determination vs. compression ratio for 10 comparison scenarios and 14 compression ratios.**

Summarizing the probability data by averaging over all comparison scenarios shows a slight increasing trend with increasing compression ratio as shown in Figure 13. This increase, however, is less than the variation among the compression levels. Moreover, inspection of Figure 12 shows that the incidence of the inconclusive determination is more likely with comparisons of inked rolled to inked flat prints over all compression levels. Inspection of images for a number of the inconclusive pairings suggests that the inability to make the definite identification decision may have more to do with ambiguity over the region of overlap of the flat print with the rolled rather than with compression effects.

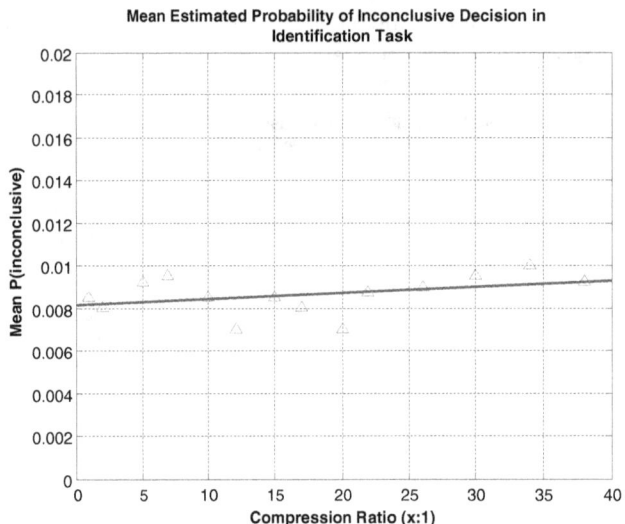

Figure 13 - Summaries over all comparison scenarios showing least squares fit to the mean probability of inconclusive determination. Slight increasing trend is noted with increasing compression.

4. Results

4.1. Investigative Goal 1: Validate 15:1 target compression ratio

The first investigative goal of this study is to validate the current 15:1 target for effective compression ratio as suggested in the informative guidance for 1000 ppi [MTR1] using the legacy methodology that formed the 500 ppi guidance [FITZPATRICK]. The IAI results (see Appendix A) utilized an error count threshold that was judged to yield an acceptable amount of detail loss using lossy WSQ compression for 500 ppi ink-captured rolled fingerprint imagery. Based on the IAI study, 7 of 202 image pairs compared (3.4 % of the image pairs) at 15:1 exhibiting non-Galton-level detail loss and zero image pairs exhibiting Galton-level detail loss constituted the acceptable level of degradation in establishing 15:1 compression for WSQ.

4.1.1. Investigative Analysis 1

The experimental case from the present study that maps to the IAI experimental scenario is the Ink-Rolled-to-Ink-Rolled-Match-Pair case (Table 2 - Case 1 and Figure 14 below), as the IAI study focused only on rolled fingerprint impressions. In this case, the present study notes 8 observations of non-Galton-level detail loss (level-3 feature loss) or 4 % at the 15:1 compression ratio. This falls slightly outside of the 3.4 % guidance used to establish 15:1. If similar guidance were to be used, a ratio of 10:1 would be the highest acceptable compression ratio to meet the IAI criterion for the rate of non-Galton-level detail loss.

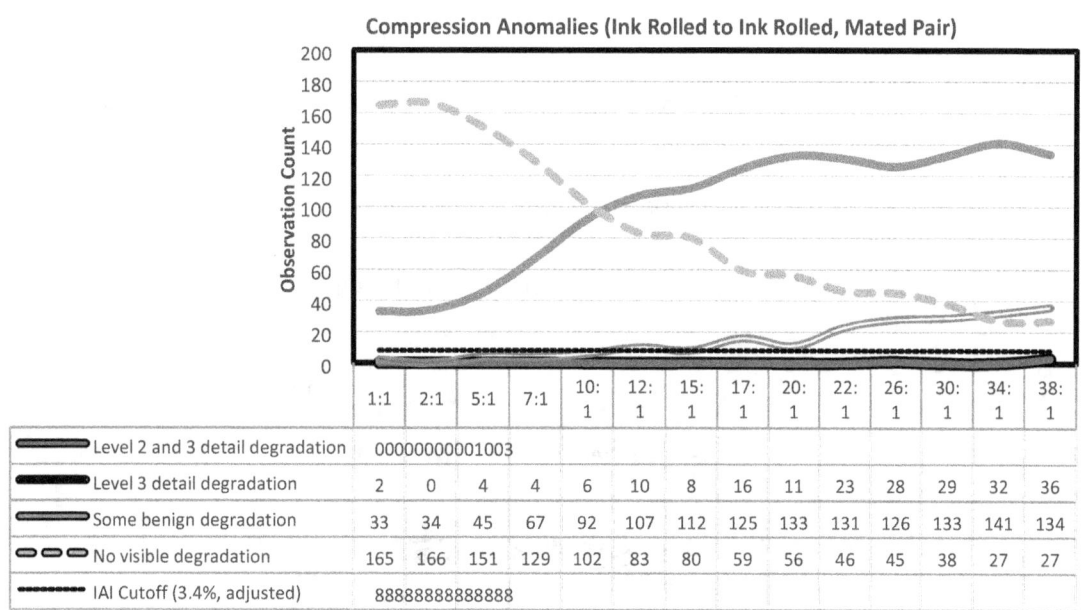

Figure 14 - Observed Compression Anomalies for Case-1 (Ink Rolled to Ink Rolled)

4.1.2. Investigative Result 1

The results of the study show that 15:1 compression of rolled-to-rolled ink card scan imagery at 1000 ppi using the current informative guidance [MTR1] falls just slightly outside of the same criteria used to establish this compression guidance at 500 ppi. While 15:1 compression of 1000 ppi imagery using the current informative guidance is a viable compression ratio and does not result in significant errors relative to lower compression ratios (see section 0), a ratio of 10:1 would meet the legacy IAI requirement while at the same time not resulting in any significant change in identification error rates as a result of compression.

4.2. Investigative Goal 2: Examine image degradation relative to compression ratio

The second investigative goal of this study was to assess if higher compression ratios result in increased perceived image degradation, and note any patterns in degradation relative to compression.

4.2.1. Investigative Analysis 2

Examination of computed Z-scores (see Table 8) yields trends that indicate increased perceived image degradation with progressively higher compression rates relative to the mean at 1:1 (control case/no compression). Such negative sloping trends can be observed in the Z-score linear trend graphs for both ink card scan (Figure 15) and digital live capture (Figure 16) cases. All cases trend negatively up to the maximum compression ratio examined in this study (38:1). The actual data from which these linear trends were generated from can be found in Figure 18 and Figure 19.

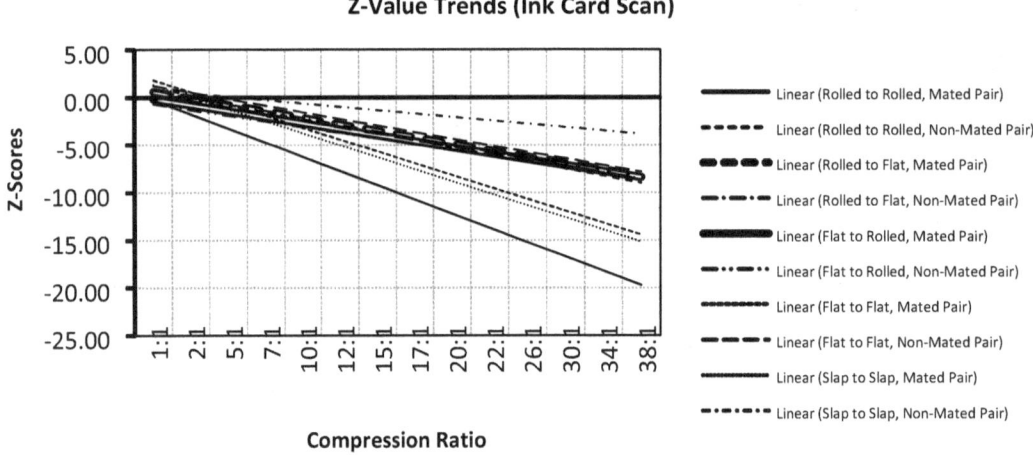

Figure 15 - Z-Score Trends (linear) for Ink Card Scan

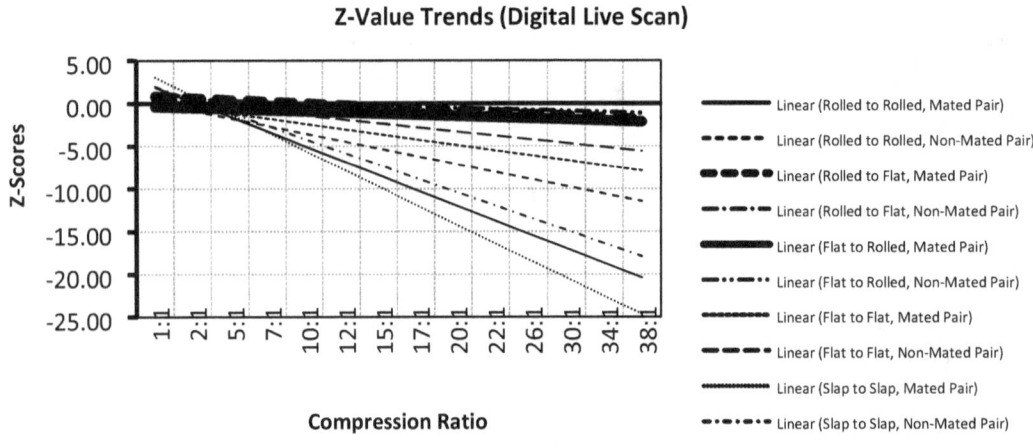

Figure 16 - Z-Scores Trends (linear) for Digital Live Capture

4.2.2. Investigative Result 2

Examination of standard scores (Z-scores) relative to the mean demonstrates that perceived image quality trends negatively with increased compression across all image types as shown by Figure 15 and Figure 16 demonstrating that increasing compression rates yields images that are perceived as having progressively greater amounts of image degradation relative to the non-compressed image as judged by human examiners. Examination of the degradation trends also show that certain impression types are impacted more than others. For example, the ink card scan rolled-to-rolled-match-pair case degrades more with increasing compression than any of the other ink card scan impression types. The same trending can be seen with digital live scan slap-to-slap-match-pair relative to other digital live scan impression types.

4.3. Investigative Goal 3: Assess impact of compression on identification error rates

The third investigative goal of this study was to assess if higher compression ratios result in increased identification error rates for fingerprint examiners, and note any patterns in the error rates.

4.3.1. Investigative Analysis 3

Analysis of observation data shows that identification errors for the experimental data occurred at very low frequency and only for a very small subset of image impression types. Specifically, only rolled-to-flat (or vice versa) image comparisons for both ink card scan and digital live capture images exhibited any errors in observation. Analysis of the observed error rates in section 3.2.1 showed that no case posed any statistically significant source of error at any compression ratio, up to and including the highest compression ratio in this study which was 38:1.

This reinforces the adage of "driving a car with a dirty windshield" where the operator of the vehicle may observe and identify severe anomalies in the field of view, but these anomalies still will not impact the operator's ability to control the vehicle and navigate the road.

It is also theorized that the features needed for most identification cases reside in the lower end of the frequency spectrum rendering such core features highly compressible, while non-Galton features may be occupying higher frequency bands and contributing the most to image entropy rendering them more costly in terms of compression. For the purposes of discussion, Figure 17 below shows examples of images compressed at up to 800:1 which shows that the ridge structure of the example set remains mostly intact even at 400:1 compression.

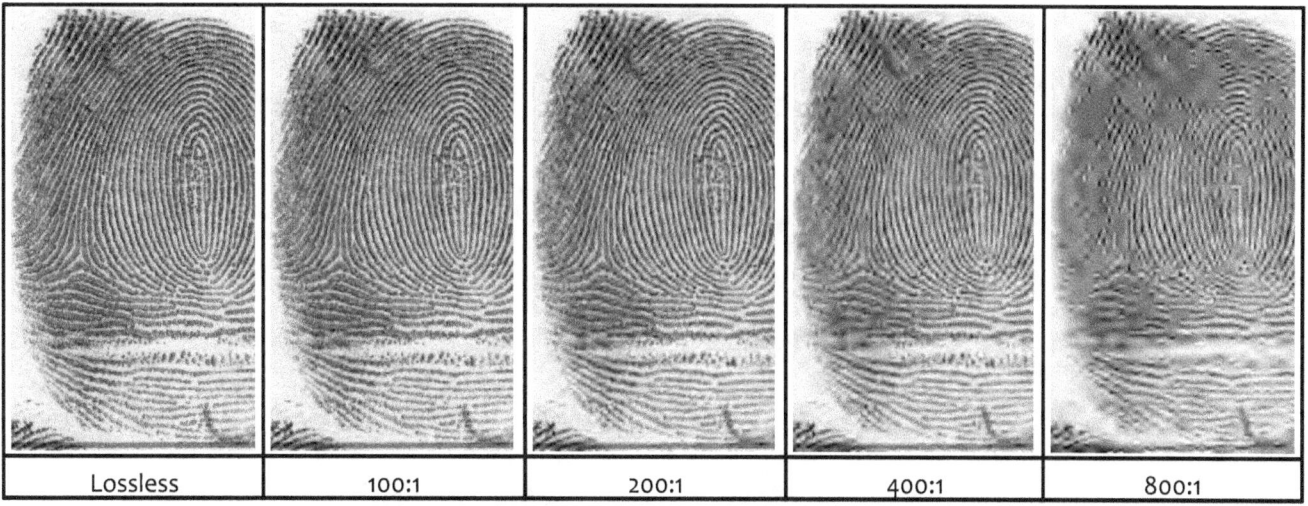

Figure 17 - Comparison of Very High Compression Rates

4.3.2. Investigative Result 3

Analysis of the observed error rates in section 3.2 demonstrates that while expert fingerprint examiners can identify image degradation and feature loss at relatively low compression rates, their ability to make an identification does not appear to be measurably impacted. This seemed to hold true even up to the maximum compression level examined in this study (38:1) where the rates of identification errors for each experimental case were not statistically significant when compared to the 1:1 control pairing (see section 0). However, given the opinion of the examiners that images with increasingly greater compression exhibited sufficient degradation to the extent that it may impact identification, it is plausible to conclude that systematic use of images with very high compression is inadvisable even if a statistically significant impact to identification rates is not observed with the limited data sample set utilized in this study. An analogy that describes this can be observing for cracks in a structural component of a bridge under increasing load where the cracks represent image feature degradation and increasing load represents increasing compression. As load is increased, a greater number of cracks are being observed. Observing an increasing number of cracks is a negative indicator and a disincentive for using very high loads even if the bridge doesn't fail under the given set of test conditions.

The disincentive for utilizing very high compression may also lie in the examination of latent fingerprint matching where features beyond traditional features may play a key role in establishing identity. While very high compression rates may yield images that can be utilized for establishing identity in 10-print casework, their utilization in latent case work may be greatly impacted.

4.4. Investigative Goal 4: Examine compression anomalies relative to impression type

The fourth investigative goal of this study was to determine if any particular fingerprint impression type is more susceptible to compression related anomalies than other impression types at the various examined compression ratios.

4.4.1. Investigative Analysis 4

As noted in section 4.2, examination of Z-scores for all cases shows that all cases trend negatively in terms of perceived image quality with increasing compression rates. Examination of the actual data plots shows some stratification of the data that indicates that not all impression types are impacted equally. This stratification can be seen in Figure 18 and Figure 19 below.

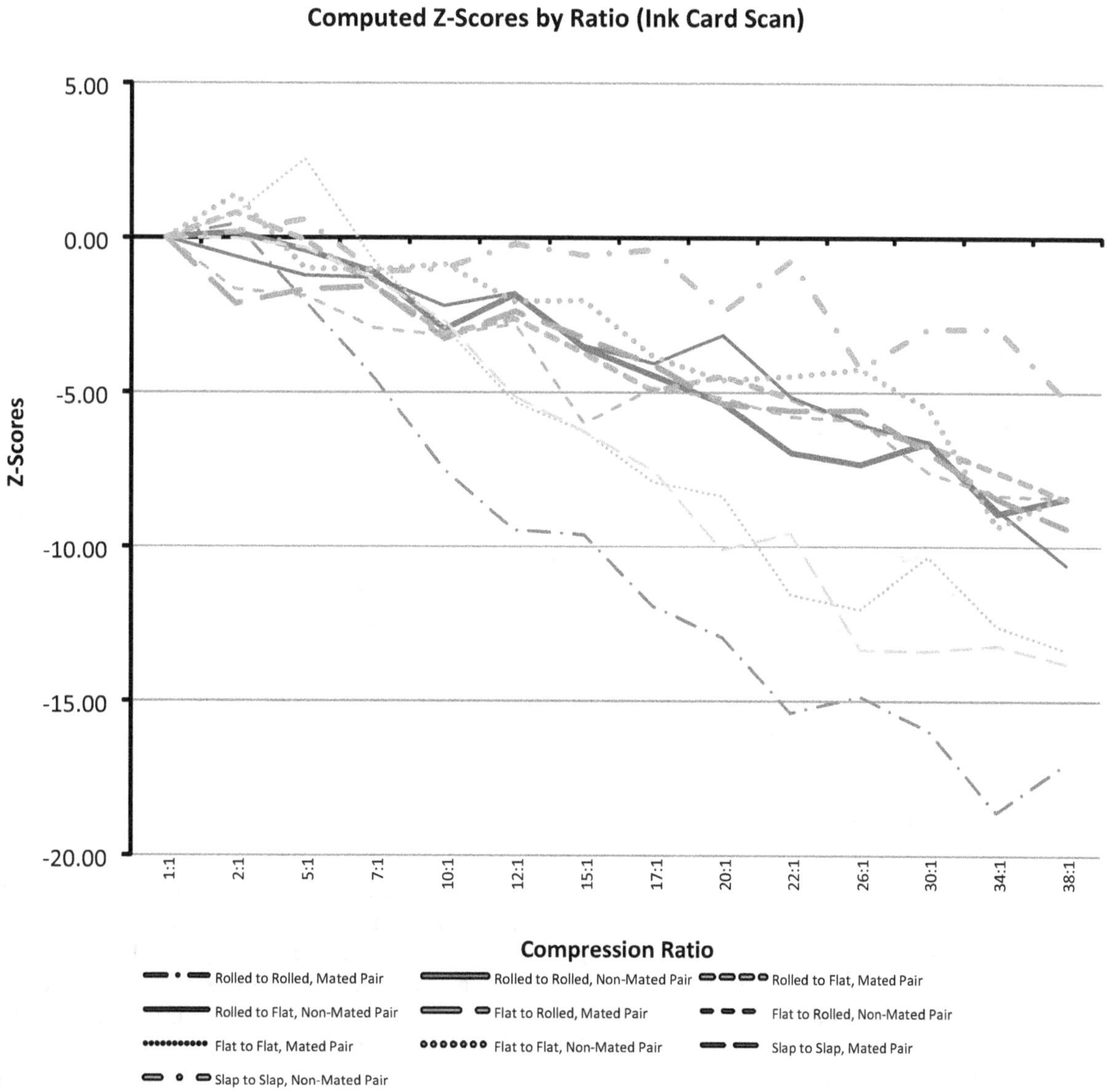

Computed Z-Scores by Ratio (Ink Card Scan)

Figure 18 - Z-Scores for Ink Card Scan Cases

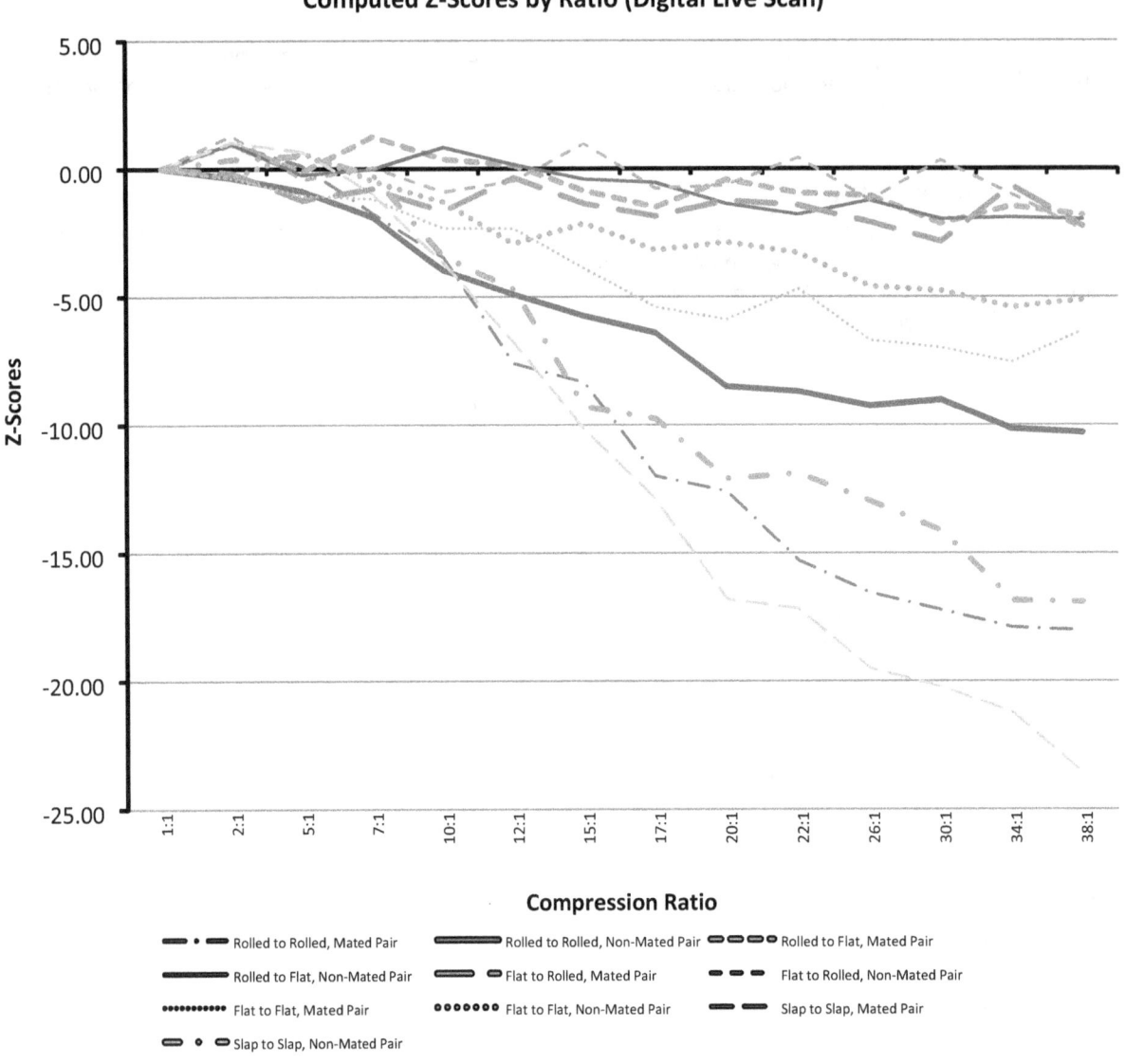

Figure 19 - Z-Scores for Digital Live Capture Cases

4.4.2. Investigative Result 4

Some impression types are more susceptible to compression degradation such as the ink card scan rolled-to-rolled case which exhibits a far more aggressive negative trend in Z-scores than the ink card scan flat- to-flat case. Not all of these cases present statistically significant results at very low compression ratios (see Table 31 and Table 32 in Appendix E) as compared to their non-compressed counterparts, but cases such as the aforementioned ink card scan rolled-to-rolled vs. ink card scan flat-to -flat case exhibit statistically significant results at compression levels as low as 5:1. For example, the rolled-to-rolled ink card scan case shows significant degradation over the mean with a Z-score of -2.068 at 5:1 while the flat-to-flat ink card scan case shows a significant improvement over mean with a Z-score of 2.530 at 5:1, both at $p<0.05$ level.

As noted above in the case of flat-to-flat ink card scan at 5:1, an interesting finding in this study was that certain cases and/or certain image types at various levels of lossy compression levels exhibit a positive trend in quality relative to the non-compressed case in terms of perceived image quality as noted by the examiners. Based on anecdotal evidence as well as some empirical studies of image entropy, it is theorized that the perceived improvement may be due to some low-pass filtering/noise removal effect on the given image. The Z-scores do provide supporting evidence that this is indeed the case. Examination of the Z-scores in Table 8 demonstrates that 13 of the 20 cases exhibit a standard score above the mean (Z-score >0), and of these 13 cases 12 of them do so at the ratio of 2:1 indicating that the image has been perceived as having better quality than the mean compared to the 1:1 control case. The evidence for this is not general nor statistically significant for most cases except one case, ink card scan flat-to-flat match pair, at the compression ratio of 5:1 (satisfying the condition of $z>0$, $p<0.05$).

5. Conclusions

This experiment was conducted with four primary goals:

1. Validate the 15:1 target compression ratio as defined in the current informative guidance for 1000 ppi using the legacy methodology that formed the 500 ppi guidance.
2. Assess if higher compression ratios result in increased perceived image degradation.
3. Assess if higher compression ratios result in increased identification error rates for fingerprint examiners.
4. Assess if any particular fingerprint image type is more susceptible to compression related anomalies.

Based on the results, it can be concluded that the existing informative guidance [MTR] for 15:1, while viable for usage in identification without resulting in any significant amount of error in identification decisions, may fall just outside of the criteria used by the IAI in establishing the effective compression ratio for 500 ppi. A lower effective compression ratio can be more helpful in the retention of non-Galton level details which may play an important role for more difficult operational scenarios such as latent fingerprint images.

The study also showed that perceived image quality does trend negatively with increased compression across all impression type comparisons. While this effect varies according to the type of image pairing being examined, the trends were all consistently negative with increasing compression.

Finally, the study shows that compression levels up to the maximum compression ratio of 38:1 examined in this study did not result in any significant impact on the ability of the examiners to make their identification decision thereby reinforcing the anecdotal evidence that human examiners can mitigate quite a bit of image quality loss while still successfully conducting the identification decision.

6. Future Work

The work laid out in this study is informative, but will serve as the basis for a normative guidance on compression for 1000 ppi fingerprint imagery. This normative guidance will be published separately by NIST, and will also take into account data from other related studies.

Also, given that compression impacts some impression types more than others, it may be possible to adopt a compression strategy that incorporates the image impression type into the decision making pathway and dynamically adjust compression.

References

Publications and Reports

CHAMBERS	Chambers, John; William Cleveland, Beat Kleiner, and Paul Tukey (1983). Graphical Methods for Data Analysis. Wadsworth
CIA1	"The World Factbook", https://www.cia.gov/library/publications/the-world-factbook/geos/us.html, Retrieved 2010-12-06.
DOJ	"The Science of Fingerprints" [rev. December 1984], United States Department of Justice, Federal Bureau of Investigation, USGPO, ISBN 0-16-050541-0, Page 18.
EYPASCH	Eypasch, Ernst; Rolf Lefering, C K Kum, Hans Troidl (1995-09-02). "Probability of adverse events that have not yet occurred: a statistical reminder". BMJ 311 (7005): 619-620. PMID 7663258. PMC 2550668. http://www.bmj.com/cgi/content/full/311/7005/619. Retrieved 2008-04-15.
FITZPATRICK	Fitzpatrick, M. et al. 1994, "WSQ Compression / Decompression Algorithm Test Report", IAI Annual Conference.
GALTON	Galton, F. (2005). *Finger prints*. Mineola, NY: Dover Publications. (Original work published 1892)
HANLEY	Hanley, JA; Lippman-Hand A (1983). "If nothing goes wrong, is everything alright?". JAMA 249 (13): 1743-5. PMID 6827763.
JAIN	Jain, A., "Pores and Ridges: High-Resolution Fingerprint Matching Using Level 3 Features", IEEE Transactions on Pattern Analysis and Machine Intelligence, Vol. 29, No. 1, January 2007.
JOVANOVIV	Jovanoviv, B. D., and P. S. Levy. A look at the Rule of Three. The American Statistician, Vol. 51, No. 2, May 1997, pp. 137-139.
LIBERT	"A 1D Spectral Image Validation/Verification Metric for Fingerprints". Libert, J.M.; Grantham, J.; Orandi, S. August 19, 2009. http://www.nist.gov/customcf/get_pdf.cfm?pub_id=903078. Retrieved 2011-01-12.
LIKERT	Likert, R. (1932). *A Technique for the Measurement of Attitudes*, Archives of Psychology 140, 55.
MTR1	"Profile for 1000 ppi Fingerprint Compression". Lepley, M.A. http://www.mitre.org/work/tech_papers/tech_papers_04/lepley_fingerprint/lepley_fingerprint.pdf. Retrieved 2011-01-11.
NIST1	National Institute of Standards and Technology. *Summary of NIST Patriot Act Recommendations*. Gaithersburg, MD. Retrieved January 4, 2007 from http://www.itl.nist.gov/iad/894.03/pact/NIST_PACT_REC.pdf
NIST2	"NIST Biometric Image Software". http://Fingerprint.nist.gov/NFIS/. Retrieved 2011-01-12.
OPENJPEG	"OpenJPEG library : an open source JPEG 2000 codec". http://www.openjpeg.org/index.php?menu=news. Retrieved 2011-01-12.
SD27	M.D. Garris & R.M. McCabe, "NIST Special Database 27: Fingerprint Minutiae from Latent and Matching Tenprint Images," NIST Technical Report NISTIR 6534 & CD-ROM, June 2000.
SHAPIRO	Shapiro, S. S.; Wilk, M. B. (1965). "An analysis of variance test for normality (complete samples)". *Biometrika* 52 (3-4): 591–611. doi:10.1093/biomet/52.3-4.591. JSTOR 2333709 MR205384.
WU1	Wu, Jin Chu, Alvin F. Martin, and Raghu N. Kacker. Measures, uncertainties, and significance test in operational ROC analysis, Journal of Research of the National Institute of Standards and Technology, 116(1), 517-537, (2011).
WU2	Wu, Jin Chu. Studies of Operational Measurement of ROC Curve on Large Fingerprint Data Sets Using Two-Sample Bootstrap. NISTIR 7449, U.S. Department of Commerce, National Institute of Standards and Technology, September 2007, 25 pages.
WU3	Wu, Jin Chu. Operational Measures and Accuracies of ROC Curve on Large Fingerprint Data Sets. NISTIR 7495, U.S. Department of Commerce, National Institute of Standards and Technology, May 2008, 23 pages.

Standards

AN27	NIST Special Publication 500-271: American National Standard for Information Systems — *Data Format for the Interchange of Fingerprint, Facial, & Other Biometric Information – Part 1.* (ANSI/NIST ITL 1-2007). Approved April 20, 2007.
JPEG	"T.81 : Information technology – Digital compression and coding of continuous-tone still images – Requirements and guidelines". http://www.itu.int/rec/T-REC-T.81. Retrieved 2011-01-12.
JPEG2K	"ISO/IEC 15444-1:2004 - Information technology -- JPEG 2000 image coding system: Core coding system". http://www.iso.org/iso/iso_catalogue/catalogue_ics/catalogue_detail_ics.htm?csnumber=27687. Retrieved 2009-11-01.
WSQ	"WSQ Gray-Scale Fingerprint Image Compression Specification" Version 3.1. https://www.fbibiospecs.org/docs/WSQ_Gray-scale_Specification_Version_3_1.pdf. Retrieved 2010-01-11.

Appendix A. IAI WSQ Compression / Decompression Study Summary

The study that became the normative basis for the 15:1 compression guidance currently in use is the IAI WSQ Compression Decompression Algorithm Test Report [FITZPATRICK]. In the IAI study, 100 fingerprint cards were selected from the Illinois State Police AFIS system. Four fingerprints were identified from each card and scanned in as 500 ppi 8bit gray-scale images. Each one of the four fingerprints were randomly compressed at settings of 5:1, 10:1, 15:1, and 20:1 totaling 400 rolled fingerprints. The selection process yielded approximately 100 fingerprints at each of the compression settings. The fingerprints were then decompressed and the resulting image paired (split-screen fashion) with the original and printed together on a single sheet with the left image consistently being the non-compressed image and the right image being the image that has passed through WSQ compression at a given ratio, but the ratio was not identified on the printed image. The selection process employed for the cards yielded a candidate pool that was comprised of 86 % males, and 14 % females. The pattern classes were balanced to reflect a distribution of 65 % loops, 29 % whorls and 6 % arches. The quality of the test fingerprints was identified as ranging from poor to good but no specific metrics were specified.

Two experienced and competent latent print examiners were each provided with the full test set of 400 paired [printed[8]] images. Each examiner independently compared the standard image (non-compressed) with the compressed/decompressed image to determine any differences between the standard image and the compressed/decompressed Image. Each examiner recorded his findings on the evaluation form by marking one of the three following evaluation codes for each test pairing:
1. No noticeable reduction in image quality
2. Slight reduction in image quality which may interfere with an identification based on poroscopy, ridgeology, or other non-Galton details.
3. Noticeable reduction in image quality which may interfere with an identification based on the Galton details.

Only one evaluation code was assessed for each fingerprint evaluated. If a test pairing contained a combination of Code 2 and Code 3 observations, Code 3 was used to evaluate the test pairing. When an evaluation Code 3 was assigned to a test pairing, the examiner highlighted the minutiae area on the test pairing and recorded the observation on the test evaluation form. When an evaluation Code 2 was assigned to a test pairing, the specific problem was recorded on the test evaluation form. Upon completion of the test, each latent print examiner turned over the completed test evaluation forms to the test coordinator. The test results were tabulated and a post-test review was conducted between the test participants and the test coordinator. Differences of opinions were resolved using a third examiner and the test conclusions were finalized. The results of the study are as follows:

Table 12 - IAI Study Results

Compression Level	Result Code			Total
	1	2	3	
5:1	202	0	0	202
10:1	200	0	0	200
15:1	195	7	0	202
20:1	37	159	0	196
Total	634	166	0	800

Based on the IAI results, an incidence of 3.4 % in Code 2 responses was deemed as acceptable and the compression ratio of 15:1 was recommended as the acceptable standard setting for the transmission and storage of electronic fingerprint images.

---

[8] Very little information could be recovered on the exact printing process used to produce the IAI study's examination deck. Based on interviews with individuals who were somewhat familiar with the original study, it appears that the images were printed using a specialized high-resolution thermal printer on specialized thermal paper in order to minimize any aberrations caused by the printing process.

The following table summarizes the key differences between the protocols of the IAI study and this study.

Table 13 - Key protocol differences between IAI WSQ study and This Study

	IAI WSQ Study	This Study
Examiners	Two	Three
Deadlock Resolution	Third examiner for deadlocked cases	Not needed
Capture Types Examined	Ink card scan	Ink card scan and digital live scan
Impression Types Examined	Rolled prints only	Rolled, Flat and Slap prints
Compression ratios examined	5:1, 10:1, 15:1 and 20:1	2:1, 5:1, 7:1, 10:1, 12:1, 15:1, 17:1, 20:1, 22:1, 26:1, 30:1, 34:1, and 38:1
Impression pairings examined	Rolled to Rolled only.	Rolled to Rolled, Rolled to Flat, Flat to Rolled, Flat to Flat and Slap to Slap (see Table 2)
Identification determination requested for Pairing	No	Yes

Appendix B. **Dataset Makeup**

For the ink card scan portion of the tests, the study utilized fingerprint images based on the Base Demonstration Model (BDM) fingerprints utilized in early tests of the FBI IAFIS system, and later used as the basis for the NIST SD-27 special database [SD27]. This ink card scan data was collected as a result of law enforcement activities and represents actual field data with collection dates ranging from 08/18/1973 through 04/12/1994. The original FD-249 fingerprint collection cards with these images were retrieved by NIST and rescanned at 1000 ppi by NIST personnel under controlled conditions. The images were scanned at 8 bits per pixel gray-scale using FBI certified software (Appendix F complaint) and stored in a non-compressed format to ensure no compression anomalies are introduced into the original set.

For the digital live capture portion of the tests, the study again utilized actual operational data captured during normal enforcement activities with collection dates ranging from 01/04/2010 through 04/13/2010. The digital live capture data was stored in an non-compressed format to ensure no compression anomalies are introduced into the original set prior to inclusion in the compression study.

Where possible, the image sets were equally balanced by gender, finger, pattern class and hand. It should be noted that balancing equally does not follow the natural demographic behavior of the population such as gender (48 % males/52 % female [CIA1]) or pattern class (65 % Loops, 30 % Whorls, 5 % Arches [DOJ]). The goal in having equal distributions of each subsample was to present a sufficiently large set of each subsample so that they are equally represented to the examiners and avoid the potential statistical bias of very small subsamples. That is, all subsamples are equally important with respect to compression irrespective of their relative incidence in the population.

Selection of match and non-match data for ink card scan and digital live-scan sets were made from each respective set and pairing between the two was not made. In comparing a particular impression to that same impression, the same exact image/impression was used. For example, in the rolled-to-rolled match cases, one rolled fingerprint image served as both the original uncompressed image as well as the compressed case image (see Case 1 in Figure 20). In the cases where one impression could not be used (i.e., comparing a flat impression to a rolled impression, such as Case 2 in Figure 20) different impressions from the same person were used in formulating the pair. For non-match data, an image from a similar pattern class was selected but from a different subject.

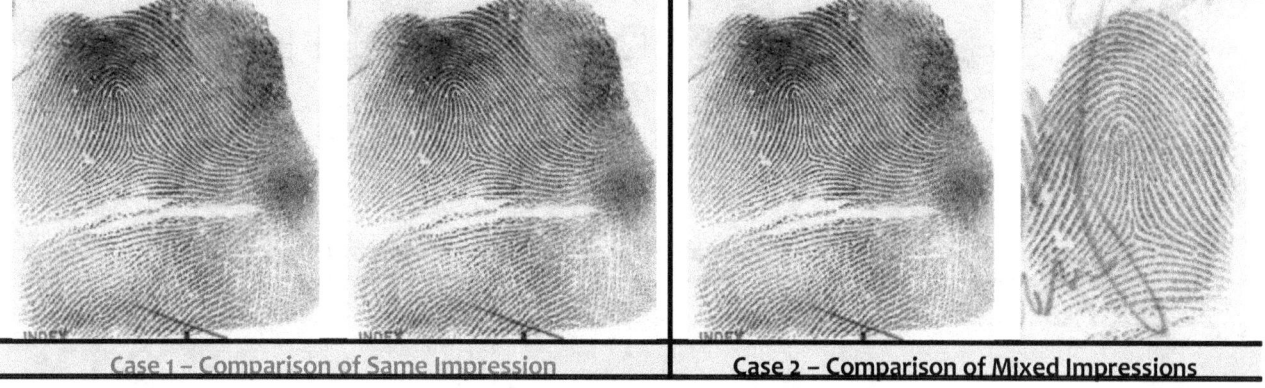

Figure 20 - Impression Comparison Examples

Demographic Make-up of Ink Card Scan Datasets

Ink card scan images used in this study consisted of 200 rolled images, 200 flat images, and 200 four-finger slap impressions.

Table 14 - Ink Card Scan Data classification by Impression Type

All Data				
Impression Type	Males	Females	Right	Left
Flat Single Finger	100	100	96	104
Rolled Single Finger	100	100	96	104
Four Finger Slaps	100	100	100	100

Table 15 - Ink Card Scan Pattern Classification for Single Finger Images by Impression Type

Data From Females (Single Finger)						Data From Males (Single Finger)				
Pattern Class	Flat	Rolled	Right	Left		Pattern Class	Flat	Rolled	Right	Left
Arch	34	34	30	38		Arch	34	34	34	34
Loop	33	33	32	34		Loop	33	33	32	34
Whorl	33	33	34	32		Whorl	33	33	34	32
Total	100	100	96	104		Total	100	100	100	100

Table 16 - Ink Card Scan Pattern Classification for Single Finger Images by Finger (Females)

Data From Females (Single Finger)											
Pattern Class	R. Thumb	R. Index	R. Middle	R. Ring	R. Little	L. Thumb	L. Index	L. Middle	L. Ring	L. Little	Total
Arch	4	4	3	4	0	3	5	4	3	4	34
Loop	3	3	4	3	3	4	3	3	4	3	33
Whorl	3	4	3	3	4	3	3	4	3	3	33
Total	10	11	10	10	7	10	11	11	10	10	100

Table 17 - Ink Card Scan Pattern Classification for Single Finger Images by Finger (Males)

Data From Males (Single Finger)											
Pattern Class	R. Thumb	R. Index	R. Middle	R. Ring	R. Little	L. Thumb	L. Index	L. Middle	L. Ring	L. Little	Total
Arch	4	3	3	4	3	3	4	3	3	4	34
Loop	3	3	4	3	3	4	3	3	4	3	33
Whorl	3	4	3	3	4	3	3	4	3	3	33
Total	10	10	10	10	10	10	10	10	10	10	100

Make-up of the digital live capture data sets

Digital live capture images used in this study consisted of 200 rolled images, 200 flat images, and 200 four-finger slap impressions.

Table 18 - Live-Scan Data classification by Impression Type

All Data Records				
Impression Type	Males	Females	Right	Left
Flat Single Finger	100	100	100	100
Rolled Single Finger	100	100	100	100
Four Finger Slaps	100	100	96	104

Table 19 - Live-Scan Pattern Classification for Single Finger Images by Impression Type

Data From Females (Single Finger)					Data From Males (Single Finger)				
Pattern Class	Flat	Rolled	Right	Left	Pattern Class	Flat	Rolled	Right	Left
Arch	33	33	36	30	Arch	33	33	36	30
Loop	34	34	30	38	Loop	34	34	30	38
Whorl	33	33	34	32	Whorl	33	33	34	32
Total	100	100	100	100	Total	100	100	100	100

Table 20 - Live-Scan Pattern Classification for Single Finger Images by Finger (Females)

Data From Females (Single Finger)										
Pattern Class	R. Thumb	R. Index	R. Middle	R. Ring	R. Little	L. Thumb	L. Index	L. Middle	L. Ring	L. Little	Total
Arch	4	4	4	3	3	3	3	3	3	3	33
Loop	3	3	3	3	3	3	4	4	4	4	34
Whorl	3	3	3	4	4	4	3	3	3	3	33
Total	10	10	10	10	10	10	10	10	10	10	100

Table 21 - Live-Scan Pattern Classification for Single Finger Images by Finger (Males)

Data From Males (Single Finger)										
Pattern Class	R. Thumb	R. Index	R. Middle	R. Ring	R. Little	L. Thumb	L. Index	L. Middle	L. Ring	L. Little	Total
Arch	4	4	4	3	3	3	3	3	3	3	33
Loop	3	3	3	3	3	3	4	4	4	4	34
Whorl	3	3	3	4	4	4	3	3	3	3	33
Total	10	10	10	10	10	10	10	10	10	10	100

Data demographics

The fingerprint images used to compile the datasets as described above were taken from several subjects. The balancing of the samples used was based on the uniqueness of a single fingerprint and not individual subjects. As such, multiple, yet distinct, fingerprint impressions were taken from some subjects (i.e., some subjects contributed more than one finger).

Table 22 - Gender Breakdown for Data

Subjects by Gender and Race	Males	Females	White	Black	Hispanic	Asian
Ink Capture Dataset	72	17	38	47	3	1
Live Capture Dataset	60	63	11	51	59	2
Combined Dataset (ALL)	132	80	49	98	62	3

Table 23 - Age Breakdown for Data

Subjects by Age	Under 18	18-24	25-29	30-34	35-39	40-44	45-49	50+
Ink Card Scan Dataset	5	54	18	3	1	4	3	1
Digital Live Capture Dataset	9	38	23	15	18	8	4	8
Combined Dataset (ALL)	14	92	41	18	19	12	7	9

Table 24 - Other Metadata: Height and Weight

Subjects by Height and Weight	<5'0"	5'0" – 5'5"	5'6" – 5'11"	6'0"+	<100 lbs	100 – 149 lbs	150 – 199 lbs	200 – 249 lbs	250+ lbs
Ink Card Scan Dataset	0	11	59	19	0	27	48	10	4
Digital Live Capture Dataset	4	45	63	11	1	45	56	17	4
Combined Dataset (All)	4	56	122	30	1	72	104	27	8

Table 25 - Other Metadata: Eye Color

Subjects by Eye Color	Brown	Black	Blue	Green	Hazel
Ink Card Scan Dataset	69	2	12	1	5
Digital Live Capture Dataset	111	1	2	6	3
Combined Dataset (ALL)	180	3	14	7	8

Table 26 - Image Geometry Data

Impression Type	Image Width (Pixels)				Image Height (Pixels)				Image Size (KB, Non-compressed)			
	Mean	Median	Min	Max	Mean	Median	Min	Max	Mean	Median	Min	Max
Ink Card Scan Rolled	1016	1002	652	1718	1166	1165	643	2063	1170.8	1134.4	535.0	2568.6
Ink Card Scan Flat	602	592	444	843	785	801	497	1008	459.6	476.0	269.0	564.1
Ink Card Scan Slap	3192	3192	3045	3334	2009	2013	1744	2082	6264.4	6264.7	5339.3	6597.5
Digital Live Scan Rolled	1600	1600	1600	1600	1500	1500	1500	1500	2343.8	2343.8	2343.8	2343.8
Digital Live Scan Flat	687	674	515	825	1057	993	672	1500	725.1	652.5	377.9	1171.9
Digital Live Scan Slap	3200	3200	3200	3200	2000	2000	2000	2000	6250.0	6250.0	6250.0	6250.0

Appendix C. Equipment Used for Study

1 x Commodity Router:
The router provides connectivity among the three workstations and NAS device, as well as providing remote access to the workstations and NAS for administration.

3 x Commodity Workstations:
The workstations are configured with 8 GB RAM, 300 GB HD, 64-bit operating system, FIXT software and data.

3 x 24 inch Monitors:
The IPS-panel monitors are connected via DVI-D and calibrated (see below) for optimal accuracy and consistency. The monitors were operated at their native resolution of 1920x1200, yielding a dpi measurement of approximately 94.3ppi. The zoom functionality in software allowed the examiners to zoom in and out of the image, and view them in the range of approximately 10x to 50x given these specific monitors.

1 x Network Attached Storage (NAS):
The NAS contains master copies of the FIXT software and data, as well as iterative copies of each stations logs/results (saved at the end of each session).

1 x Monitor Calibration Device:
The monitors were calibrated using a system which consists of a colorimeter paired with proprietary software designed specifically for use with the colorimeter and for the purpose of monitor calibration. The colorimeter is a sensor which provides an accurate measurement of colors as they actually appear on the monitor screen. During the calibration process, the colorimeter is physically attached to the monitor while the software displays a series of solid colors on the screen. The colorimeter measures the actual color values displayed on the monitor and then provides these measurements to the software. The software uses these measurements to calculate the difference between the color values as they are displayed on the monitor against the true color values within the software. The software then applies configuration changes to the system in order to correct the color values displayed by the monitor, ensuring accurate color reproduction. Due to the fact that each monitor, even of the same model, performs slightly differently in terms of the accuracy of color reproduction, this process was completed independently on each of the three FIXT workstations.

Appendix D. Observation Data

Ink Capture Compression Degradation Observations

Table 27 - Ink Capture Degradation Results

	1:1	2:1	5:1	7:1	10:1	12:1	15:1	17:1	20:1	22:1	26:1	30:1	34:1	38:1
Case 1 - Ink Rolled to Ink Rolled, Match Pair														
Level 2 and 3 detail degradation	0	0	0	0	0	0	0	0	0	1	0	0	3	
Level 3 detail degradation	2	0	4	4	6	10	8	16	11	23	28	29	32	36
Some benign degradation	33	34	45	67	92	107	112	125	133	131	126	133	141	134
No visible degradation	165	166	151	129	102	83	80	59	56	46	45	38	27	27
Case 2 - Ink Rolled to Ink Rolled, Non-Match Pair														
Level 2 and 3 detail degradation	1	1	1	0	1	0	1	1	1	1	0	4	5	4
Level 3 detail degradation	27	34	25	32	45	40	46	47	61	63	80	66	76	77
Some benign degradation	96	80	105	104	97	98	105	116	97	104	93	99	100	95
No visible degradation	76	85	69	64	57	62	48	36	41	32	27	31	19	24
Case 3 - Ink Rolled to Ink Flat, Match Pair														
Level 2 and 3 detail degradation	4	1	0	3	5	4	4	2	4	4	7	6	6	13
Level 3 detail degradation	29	26	41	41	46	48	54	62	51	61	69	72	80	79
Some benign degradation	120	128	113	120	124	118	118	118	131	123	108	111	102	97
No visible degradation	47	45	46	36	25	30	24	18	14	12	16	11	12	11
Case 4 - Ink Rolled to Ink Flat, Non-Match Pair														
Level 2 and 3 detail degradation	2	4	4	2	3	4	4	5	1	3	6	4	5	7
Level 3 detail degradation	27	32	22	29	36	36	45	41	35	50	60	57	85	96
Some benign degradation	91	82	109	102	101	90	94	108	121	109	96	112	79	76
No visible degradation	80	82	65	67	60	70	57	46	43	38	38	27	31	21
Case 5 - Ink Flat to Ink Rolled, Match Pair														
Level 2 and 3 detail degradation	4	9	6	1	6	2	5	4	4	5	4	3	4	11
Level 3 detail degradation	32	36	35	46	51	41	48	59	75	79	75	84	104	101
Some benign degradation	109	110	118	116	109	131	116	109	101	95	102	104	75	80
No visible degradation	55	45	41	37	34	26	31	28	20	21	19	9	17	8
Case 6 - Ink Flat to Ink Rolled, Non-Match Pair														
Level 2 and 3 detail degradation	3	3	2	2	2	2	3	5	1	3	4	8	2	2
Level 3 detail degradation	25	28	30	38	41	34	47	43	55	55	52	63	81	82
Some benign degradation	82	99	102	99	96	105	115	105	102	105	108	101	87	86
No visible degradation	90	70	66	61	61	59	35	47	42	37	36	28	30	30
Case 7 - Ink Flat to Ink Flat, Match Pair														
Level 2 and 3 detail degradation	1	4	0	0	0	3	1	0	1	3	3	4	5	
Level 3 detail degradation	6	2	3	3	2	10	6	9	11	18	20	18	35	47
Some benign degradation	30	19	20	47	68	79	98	107	112	118	118	119	109	104
No visible degradation	163	175	177	150	130	111	93	83	77	63	59	60	52	44
Case 8 - Ink Flat to Ink Flat, Non-Match Pair														
Level 2 and 3 detail degradation	2	6	4	4	2	4	6	5	6	5	7	7	15	14
Level 3 detail degradation	28	17	35	35	30	41	37	39	44	54	50	55	69	78
Some benign degradation	87	75	78	79	94	84	81	105	101	88	85	94	89	65
No visible degradation	83	102	83	82	74	71	76	51	49	53	58	44	27	43
Case 9 - Ink Slap to Ink Slap, Match Pair														
Level 2 and 3 detail degradation	1	1	1	0	1	0	0	1	0	1	1	2	0	1
Level 3 detail degradation	1	0	1	4	1	9	11	11	8	8	14	14	18	24
Some benign degradation	18	20	20	26	37	51	67	80	101	103	117	116	117	115
No visible degradation	180	179	178	170	161	140	122	108	91	88	68	68	65	60
Case 10 - Ink Slap to Ink Slap, Non-Match Pair														
Level 2 and 3 detail degradation	1	0	0	0	1	1	0	1	1	0	1	0	1	2
Level 3 detail degradation	30	35	33	43	41	36	36	34	41	36	58	55	45	62
Some benign degradation	92	83	82	85	84	83	92	89	104	94	92	90	105	102
No visible degradation	77	82	85	72	74	80	72	76	54	70	49	55	49	34

Live Capture Compression Degradation Observations

Table 28 - Live Capture Degradation Results

	1:1	2:1	5:1	7:1	10:1	12:1	15:1	17:1	20:1	22:1	26:1	30:1	34:1	38:1
Case 11 - Digital Live Capture Rolled to Digital Live Capture Rolled, Match Pair														
Level 2 and 3 detail degradation	1	0	0	1	0	2	1	1	1	1	1	0	1	1
Level 3 detail degradation	2	1	1	0	2	4	2	12	13	33	45	62	63	69
Some benign degradation	38	36	43	58	75	109	124	139	146	136	131	119	117	113
No visible degradation	159	163	156	141	123	85	73	48	40	30	23	19	19	17
Case 12 - Digital Live Capture Rolled to Digital Live Capture Rolled, Non-Match Pair														
Level 2 and 3 detail degradation	3	2	5	4	3	4	2	4	6	6	5	3	5	6
Level 3 detail degradation	17	12	7	6	22	19	29	32	57	50	73	82	89	81
Some benign degradation	113	131	135	150	149	156	153	151	125	135	110	107	97	105
No visible degradation	67	55	53	40	26	21	16	13	12	9	12	8	9	8
Case 13 - Digital Live Capture Rolled to Digital Live Capture Flat, Match Pair														
Level 2 and 3 detail degradation	7	7	5	7	7	8	7	4	6	3	5	6	5	5
Level 3 detail degradation	23	13	22	14	17	12	21	29	27	30	31	34	30	42
Some benign degradation	113	120	125	116	120	130	128	132	114	127	120	123	126	108
No visible degradation	57	60	48	63	56	50	44	35	53	40	44	37	39	45
Case 14 - Digital Live Capture Rolled to Digital Live Capture Flat, Non-Match Pair														
Level 2 and 3 detail degradation	8	6	11	8	6	4	9	5	7	4	6	4	4	6
Level 3 detail degradation	20	14	12	23	18	18	21	33	22	43	32	46	50	35
Some benign degradation	121	129	128	115	122	139	120	114	138	115	121	110	105	127
No visible degradation	51	51	49	54	54	39	50	48	33	38	41	40	41	32
Case 15 - Digital Live Capture Flat to Digital Live Capture Rolled, Match Pair														
Level 2 and 3 detail degradation	5	6	5	6	6	6	4	9	6	9	11	9	4	8
Level 3 detail degradation	16	11	20	17	20	10	19	24	19	19	22	28	19	28
Some benign degradation	117	125	124	122	127	130	132	110	121	115	112	116	123	113
No visible degradation	62	58	51	55	47	54	45	57	54	57	55	47	54	51
Case 16 - Digital Live Capture Flat to Digital Live Capture Rolled, Non-Match Pair														
Level 2 and 3 detail degradation	8	7	9	9	11	8	7	8	7	8	7	7	10	9
Level 3 detail degradation	17	12	21	19	16	15	17	27	19	18	26	31	18	31
Some benign degradation	123	121	117	114	125	133	115	113	130	115	124	115	126	123
No visible degradation	52	60	53	58	48	44	61	52	44	59	43	52	46	37
Case 17 - Digital Live Capture Flat to Digital Live Capture Flat, Match Pair														
Level 2 and 3 detail degradation	1	0	1	1	1	1	0	1	0	1	1	0	1	1
Level 3 detail degradation	0	0	0	1	0	2	0	1	3	3	6	6	8	13
Some benign degradation	25	31	30	31	42	37	55	64	68	56	75	73	77	63
No visible degradation	174	169	169	167	157	160	145	134	129	140	119	120	114	124
Case 18 - Digital Live Capture Flat to Digital Live Capture Flat, Non-Match Pair														
Level 2 and 3 detail degradation	2	2	1	2	1	3	2	3	2	3	3	2	3	2
Level 3 detail degradation	2	3	5	3	4	4	5	5	7	8	5	8	10	12
Some benign degradation	79	80	72	81	91	96	89	98	92	92	108	107	111	107
No visible degradation	117	115	122	114	104	97	104	94	99	97	84	83	76	79
Case 19 - Digital Live Capture Slap to Digital Live Capture Slap, Match Pair														
Level 2 and 3 detail degradation	0	0	0	0	0	0	0	0	0	0	0	0	0	0
Level 3 detail degradation	0	0	0	0	0	0	1	2	1	3	1	3	9	11
Some benign degradation	8	5	6	12	30	50	81	103	125	129	141	143	146	149
No visible degradation	192	195	194	188	170	150	118	95	74	68	58	54	45	40
Case 20 - Digital Live Capture Slap to Digital Live Capture Slap, Non-Match Pair														
Level 2 and 3 detail degradation	0	0	0	0	0	0	0	0	0	0	0	0	0	0
Level 3 detail degradation	2	1	0	0	1	3	2	3	4	3	8	10	8	17
Some benign degradation	38	37	38	44	70	78	124	126	140	144	146	149	165	152
No visible degradation	160	162	162	156	129	119	74	71	56	53	46	41	27	31

Ink Capture Compression Identification Observations

Table 29 - Ink Capture Identification Results

	1:1	2:1	5:1	7:1	10:1	12:1	15:1	17:1	20:1	22:1	26:1	30:1	34:1	38:1
Ink Rolled to Ink Rolled, Match Pair														
Incorrect Identification	0	0	0	0	0	0	0	0	0	0	0	0	0	0
Indeterminate Identification	0	0	0	0	0	0	0	0	0	0	0	0	0	0
Correct Identification	200	200	200	200	200	200	200	200	200	200	200	200	200	200
Ink Rolled to Ink Rolled, Non-Match Pair														
Incorrect Identification	0	0	0	0	0	0	0	0	0	0	0	0	0	0
Indeterminate Identification	3	3	3	3	1	1	4	1	3	3	3	4	2	1
Correct Identification	197	197	197	197	199	199	196	199	197	197	197	196	198	199
Ink Rolled to Ink Flat, Match Pair														
Incorrect Identification	1	1	2	1	0	2	0	0	1	1	2	1	2	2
Indeterminate Identification	5	6	5	6	7	5	6	7	4	6	5	6	8	7
Correct Identification	194	193	193	193	193	193	194	193	195	193	193	193	190	191
Ink Rolled to Ink Flat, Non-Match Pair														
Incorrect Identification	0	0	0	0	0	0	0	0	0	0	0	0	0	0
Indeterminate Identification	5	3	4	0	3	1	3	3	3	2	3	1	1	3
Correct Identification	195	197	196	200	197	199	197	197	197	198	197	199	199	197
Ink Flat to Ink Rolled, Match Pair														
Incorrect Identification	0	1	0	1	0	1	0	1	1	0	0	1	0	0
Indeterminate Identification	6	4	6	7	7	5	6	6	6	8	7	7	7	6
Correct Identification	194	195	194	192	193	194	194	193	193	192	193	192	193	194
Ink Flat to Ink Rolled, Non-Match Pair														
Incorrect Identification	0	0	0	0	0	0	0	0	0	0	0	0	0	0
Indeterminate Identification	3	4	3	5	2	3	3	3	3	3	5	4	2	5
Correct Identification	197	196	197	195	198	197	197	197	197	197	195	196	198	195
Ink Flat to Ink Flat, Match Pair														
Incorrect Identification	0	0	0	0	0	0	0	0	0	0	0	0	0	0
Indeterminate Identification	0	0	0	0	0	0	2	1	0	0	0	0	1	1
Correct Identification	200	200	200	200	200	200	198	199	200	200	200	200	199	199
Ink Flat to Ink Flat, Non-Match Pair														
Incorrect Identification	0	0	0	0	0	0	0	0	0	0	0	0	0	0
Indeterminate Identification	2	1	3	3	1	2	1	3	2	2	3	4	5	3
Correct Identification	198	199	197	197	199	198	199	197	198	198	197	196	195	197
Ink Slap to Ink Slap, Match Pair														
Incorrect Identification	0	0	0	0	0	0	0	0	0	0	0	0	0	0
Indeterminate Identification	0	0	0	0	0	0	0	0	0	0	0	0	0	0
Correct Identification	200	200	200	200	200	200	200	200	200	200	200	200	200	200
Ink Slap to Ink Slap, Non-Match Pair														
Incorrect Identification	0	0	0	0	0	0	0	0	0	0	0	0	0	0
Indeterminate Identification	0	0	0	0	0	0	0	0	0	0	0	0	0	0
Correct Identification	200	200	200	200	200	200	200	200	200	200	200	200	200	200

Live Capture Compression Identification Observations

Table 30 - Live Capture Identification Results

	1:1	2:1	5:1	7:1	10:1	12:1	15:1	17:1	20:1	22:1	26:1	30:1	34:1	38:1
Digital Live Capture Rolled to Digital Live Capture Rolled, Match Pair														
Incorrect Identification	0	0	0	0	0	0	0	0	0	0	0	0	0	0
Indeterminate Identification	0	0	0	0	0	0	0	0	0	0	0	0	0	0
Correct Identification	200	200	200	200	200	200	200	200	200	200	200	200	200	200
Digital Live Capture Rolled to Digital Live Capture Rolled, Non-Match Pair														
Incorrect Identification	0	0	0	0	0	0	0	0	0	0	0	0	0	0
Indeterminate Identification	0	1	3	2	1	1	1	3	0	3	1	1	3	1
Correct Identification	200	199	197	198	199	199	199	197	200	197	199	199	197	199
Digital Live Capture Rolled to Digital Live Capture Flat, Match Pair														
Incorrect Identification	0	0	0	0	0	0	2	1	1	1	0	0	1	
Indeterminate Identification	3	2	3	3	3	3	3	1	1	1	1	3	4	1
Correct Identification	197	198	197	197	197	197	197	197	198	198	198	197	196	198
Digital Live Capture Rolled to Digital Live Capture Flat, Non-Match Pair														
Incorrect Identification	0	0	0	0	0	0	0	0	0	0	0	0	0	0
Indeterminate Identification	1	3	1	2	1	1	1	0	1	1	2	1	0	2
Correct Identification	199	197	199	198	199	199	199	200	199	199	198	199	200	198
Digital Live Capture Flat to Digital Live Capture Rolled, Match Pair														
Incorrect Identification	1	1	0	0	0	0	1	1	0	0	0	0	0	0
Indeterminate Identification	3	2	2	2	3	2	2	2	2	2	2	3	2	3
Correct Identification	196	197	198	198	197	198	197	197	198	198	198	197	198	197
Digital Live Capture Flat to Digital Live Capture Rolled, Non-Match Pair														
Incorrect Identification	0	0	0	0	0	0	0	0	0	0	0	0	0	0
Indeterminate Identification	2	1	3	3	4	2	1	1	2	3	3	2	3	2
Correct Identification	198	199	197	197	196	198	199	199	198	197	197	198	197	198
Digital Live Capture Flat to Digital Live Capture Flat, Match Pair														
Incorrect Identification	0	0	0	0	0	0	0	0	0	0	0	0	0	0
Indeterminate Identification	0	0	0	0	0	1	0	0	0	0	0	0	0	0
Correct Identification	200	200	200	200	200	199	200	200	200	200	200	200	200	200
Digital Live Capture Flat to Digital Live Capture Flat, Non-Match Pair														
Incorrect Identification	0	0	0	0	0	0	0	0	0	0	0	0	0	0
Indeterminate Identification	1	2	1	2	1	1	1	1	1	1	1	2	2	2
Correct Identification	199	198	199	198	199	199	199	199	199	199	199	198	198	198
Digital Live Capture Slap to Digital Live Capture Slap, Match Pair														
Incorrect Identification	0	0	0	0	0	0	0	0	0	0	0	0	0	0
Indeterminate Identification	0	0	0	0	0	0	0	0	0	0	0	0	0	0
Correct Identification	200	200	200	200	200	200	200	200	200	200	200	200	200	200
Digital Live Capture Slap to Digital Live Capture Slap, Non-Match Pair														
Incorrect Identification	0	0	0	0	0	0	0	0	0	0	0	0	0	0
Indeterminate Identification	0	0	0	0	0	0	0	0	0	0	0	0	0	0
Correct Identification	200	200	200	200	200	200	200	200	200	200	200	200	200	200

Appendix E. Statistical Parameters of Bootstrap Results and Hypothesis Testing

Table 31 and Table 32 summarize means, medians, and standard errors of the mean for bootstrap replicates of the degradation score for each of 20 matching scenarios at each of 14 compression levels. Z-tests are used to test the difference between test score values for non-compressed images and those at each of the compression levels. P-values less than 0.05 (shown below in dashed boxes) indicate differences significant at the 5 % level. It should be noted that these tests examine the degradation observed at each level of compression in contrast to that observed between two images that have not been compressed. In the "id" case, both images are identical except for the level of compression applied to one of the two images in the match pair.

For each matching scenario, significance tests reveal the compression level at which observer ratings of the relative degradation depart from that of the non-compressed case. For most of the scenarios, this occurs at 10:1 compression.

Table 31 - Distribution parameters of bootstrap replications and hypothesis tests of differences in degradation between lossless baseline and compressed images (Ink Card Scan)

Scenario			1:1	2:1	5:1	7:1	10:1	12:1	15:1	17:1	20:1	22:1	26:1	30:1	34:1	38:1
Rolled to Rolled, Match Pair		Mean	0.046	0.042	0.066	0.094	0.130	0.159	0.160	0.197	0.194	0.221	0.232	0.238	0.256	0.272
		StdErr	0.007	0.007	0.009	0.009	0.010	0.010	0.010	0.010	0.010	0.010	0.011	0.010	0.010	0.012
		Median	0.046	0.043	0.066	0.094	0.130	0.159	0.160	0.196	0.194	0.221	0.233	0.239	0.256	0.273
		z	-----	0.444	2.068	-4.559	-7.518	-9.465	-9.607	-11.942	-12.930	-15.381	-14.851	-15.916	-18.603	-17.016
		p	-----	0.657	0.039	0.000	0.000	0.000	0.000	0.000	0.000	0.000	0.000	0.000	0.000	0.000
Rolled to Rolled, Non-Match Pair		Mean	0.193	0.190	0.199	0.210	0.239	0.223	0.251	0.268	0.279	0.293	0.316	0.309	0.340	0.331
		StdErr	0.012	0.013	0.012	0.012	0.013	0.012	0.013	0.012	0.013	0.012	0.012	0.014	0.013	0.013
		Median	0.193	0.190	0.199	0.210	0.239	0.223	0.251	0.268	0.279	0.293	0.316	0.309	0.340	0.331
		z	-----	0.165	-0.400	-1.123	-2.967	-1.851	-3.571	-4.464	-5.359	-6.964	-7.341	-6.633	-8.928	-8.427
		p	-----	0.869	0.689	0.262	0.003	0.064	0.000	0.000	0.000	0.000	0.000	0.000	0.000	0.000
Rolled to Flat, Match Pair		Mean	0.242	0.230	0.244	0.267	0.295	0.287	0.303	0.312	0.311	0.326	0.342	0.349	0.358	0.383
		StdErr	0.013	0.011	0.012	0.012	0.013	0.013	0.013	0.011	0.012	0.012	0.013	0.013	0.013	0.015
		Median	0.243	0.230	0.244	0.268	0.295	0.288	0.303	0.313	0.311	0.326	0.341	0.349	0.356	0.384
		z	-----	0.800	0.072	-1.586	-3.115	-2.626	-3.755	-4.960	-4.479	-5.191	-6.056	-6.738	-7.580	-8.463
		p	-----	0.424	0.942	0.113	0.002	0.009	0.000	0.000	0.000	0.000	0.000	0.000	0.000	0.000
Rolled to Flat, Non-Match Pair		Mean	0.191	0.202	0.211	0.210	0.231	0.223	0.251	0.262	0.244	0.276	0.299	0.302	0.336	0.370
		StdErr	0.014	0.015	0.013	0.013	0.014	0.015	0.015	0.015	0.012	0.014	0.015	0.013	0.014	0.015
		Median	0.191	0.203	0.211	0.210	0.231	0.223	0.250	0.263	0.244	0.276	0.299	0.303	0.336	0.370
		z	-----	-0.623	-1.232	-1.290	-2.217	-1.794	-3.499	-4.070	-3.164	-5.153	-6.043	-6.613	-8.786	-10.596
		p	-----	0.534	0.218	0.197	0.027	0.073	0.000	0.000	0.002	0.000	0.000	0.000	0.000	0.000
Flat to Rolled, Match Pair		Mean	0.236	0.272	0.265	0.265	0.293	0.276	0.290	0.303	0.333	0.341	0.335	0.355	0.374	0.408
		StdErr	0.014	0.016	0.015	0.012	0.014	0.011	0.013	0.013	0.013	0.014	0.013	0.011	0.013	0.014
		Median	0.236	0.271	0.265	0.265	0.293	0.276	0.290	0.304	0.334	0.341	0.335	0.355	0.374	0.408
		z	-----	-2.144	-1.687	-1.593	-3.269	-2.394	-3.252	-4.120	-5.364	-5.599	-5.583	-6.950	-8.443	-9.388
		p	-----	0.032	0.092	0.111	0.001	0.017	0.001	0.000	0.000	0.000	0.000	0.000	0.000	0.000
Flat to Rolled, Non-Match Pair		Mean	0.180	0.208	0.212	0.228	0.232	0.226	0.276	0.264	0.270	0.284	0.285	0.324	0.321	0.322
		StdErr	0.014	0.014	0.013	0.013	0.014	0.013	0.013	0.015	0.013	0.014	0.014	0.015	0.013	0.013
		Median	0.180	0.208	0.211	0.229	0.233	0.226	0.276	0.264	0.270	0.284	0.285	0.324	0.321	0.323
		z	-----	-1.683	-1.889	-2.937	-3.175	-2.800	-6.020	-4.865	-5.185	-5.800	-5.940	-7.588	-8.332	-8.423
		p	-----	0.092	0.059	0.003	0.001	0.005	0.000	0.000	0.000	0.000	0.000	0.000	0.000	0.000
Flat to Flat, Match Pair		Mean	0.058	0.049	0.032	0.066	0.090	0.124	0.152	0.161	0.168	0.198	0.212	0.209	0.243	0.272
		StdErr	0.010	0.011	0.007	0.008	0.009	0.011	0.012	0.011	0.010	0.011	0.012	0.012	0.014	0.014
		Median	0.058	0.048	0.033	0.066	0.090	0.124	0.153	0.161	0.168	0.198	0.211	0.209	0.244	0.271
		z	-----	0.753	2.530	-0.794	-2.915	-5.322	-6.285	-7.909	-8.362	-11.558	-12.035	-10.333	-12.580	-13.343
		p	-----	0.451	0.011	0.427	0.004	0.000	0.000	0.000	0.000	0.000	0.000	0.000	0.000	0.000
Flat to Flat, Non-Match Pair		Mean	0.189	0.166	0.205	0.206	0.203	0.228	0.224	0.254	0.266	0.270	0.267	0.290	0.359	0.346
		StdErr	0.013	0.016	0.015	0.015	0.013	0.015	0.016	0.014	0.015	0.015	0.016	0.016	0.018	0.018
		Median	0.189	0.166	0.205	0.205	0.203	0.228	0.224	0.254	0.266	0.270	0.266	0.290	0.360	0.346
		z	-----	1.395	-0.984	-1.044	-0.849	-2.073	-2.046	-3.862	-4.609	-4.479	-4.257	-5.507	-9.363	-8.282
		p	-----	0.163	0.325	0.297	0.396	0.038	0.041	0.000	0.000	0.000	0.000	0.000	0.000	0.000
Slap to Slap, Match Pair		Mean	0.030	0.030	0.032	0.043	0.054	0.086	0.111	0.132	0.146	0.153	0.186	0.190	0.191	0.209
		StdErr	0.007	0.007	0.007	0.008	0.009	0.010	0.011	0.012	0.010	0.011	0.011	0.012	0.011	0.012
		Median	0.029	0.030	0.033	0.043	0.054	0.086	0.111	0.133	0.146	0.153	0.185	0.190	0.191	0.209
		z	-----	-0.025	-0.322	-1.349	-2.696	-5.134	-6.276	-7.544	-10.102	-9.558	-13.338	-13.373	-13.197	-13.782
		p	-----	0.980	0.748	0.177	0.007	0.000	0.000	0.000	0.000	0.000	0.000	0.000	0.000	0.000
Slap to Slap, Non-Match Pair		Mean	0.195	0.192	0.185	0.213	0.212	0.199	0.205	0.202	0.237	0.208	0.265	0.249	0.249	0.293
		StdErr	0.013	0.013	0.013	0.013	0.013	0.013	0.012	0.013	0.013	0.012	0.014	0.013	0.013	0.013
		Median	0.195	0.191	0.185	0.214	0.213	0.199	0.205	0.201	0.238	0.208	0.265	0.250	0.249	0.293
		z	-----	0.173	0.584	-1.112	-1.001	-0.206	-0.572	-0.388	-2.420	-0.728	-4.245	-2.946	-2.958	-5.252
		p	-----	0.862	0.560	0.266	0.317	0.837	0.567	0.698	0.016	0.467	0.000	0.003	0.003	0.000

Table 32 - Distribution parameters of bootstrap replications and hypothesis tests of differences in degradation between lossless baseline and compressed images (Digital Live Capture)

			1:1	2:1	5:1	7:1	10:1	12:1	15:1	17:1	20:1	22:1	26:1	30:1	34:1	38:1
Digital Live Capture	Rolled to Rolled, Match Pair	Mean	0.057	0.048	0.056	0.077	0.098	0.156	0.165	0.208	0.220	0.257	0.281	0.303	0.308	0.319
		StdErr	0.009	0.007	0.008	0.009	0.009	0.011	0.010	0.010	0.010	0.011	0.011	0.011	0.012	0.011
		Median	0.058	0.048	0.056	0.078	0.099	0.156	0.165	0.209	0.220	0.258	0.281	0.304	0.308	0.319
		z	------	0.982	0.107	-1.737	-3.453	-7.597	-8.337	-11.986	-12.584	-15.285	-16.558	-17.220	-17.896	-18.014
		p	------	0.326	0.915	0.082	0.001	0.000	0.000	0.000	0.000	0.000	0.000	0.000	0.000	0.000
	Rolled to Rolled, Non-Match Pair	Mean	0.199	0.204	0.212	0.223	0.256	0.262	0.274	0.289	0.329	0.324	0.345	0.354	0.369	0.363
		StdErr	0.012	0.011	0.012	0.011	0.011	0.011	0.010	0.010	0.013	0.012	0.013	0.011	0.012	0.012
		Median	0.199	0.204	0.211	0.223	0.256	0.263	0.274	0.289	0.329	0.324	0.345	0.354	0.369	0.364
		z	------	-0.351	-0.873	-1.887	-3.969	-4.911	-5.752	-6.410	-8.495	-8.687	-9.252	-9.020	-10.156	-10.296
		p	------	0.725	0.383	0.059	0.000	0.000	0.000	0.000	0.000	0.000	0.000	0.000	0.000	0.000
	Rolled to Flat, Match Pair	Mean	0.235	0.218	0.237	0.215	0.228	0.233	0.248	0.258	0.240	0.249	0.253	0.269	0.258	0.265
		StdErr	0.015	0.015	0.014	0.015	0.014	0.014	0.015	0.013	0.015	0.012	0.014	0.014	0.013	0.015
		Median	0.234	0.218	0.236	0.215	0.229	0.233	0.248	0.258	0.240	0.249	0.253	0.269	0.258	0.265
		z	------	0.975	0.122	1.264	0.368	0.133	-0.854	-1.498	-0.432	-0.949	-1.056	2.141	-1.459	-1.809
		p	------	0.330	0.903	0.206	0.713	0.895	0.393	0.134	0.666	0.343	0.291	0.032	0.145	0.070
	Rolled to Flat, Non-Match Pair	Mean	0.241	0.226	0.245	0.241	0.228	0.239	0.248	0.250	0.263	0.271	0.262	0.273	0.276	0.277
		StdErr	0.015	0.013	0.016	0.015	0.014	0.012	0.015	0.014	0.014	0.014	0.014	0.014	0.014	0.014
		Median	0.241	0.226	0.245	0.241	0.228	0.239	0.248	0.249	0.263	0.271	0.261	0.273	0.276	0.276
		z	------	0.962	-0.251	-0.020	0.837	0.152	-0.410	-0.541	-1.371	-1.788	-1.235	-1.968	-1.909	-1.984
		p	------	0.336	0.802	0.984	0.403	0.879	0.682	0.588	0.170	0.074	0.217	0.049	0.056	0.047
	Flat to Rolled, Match Pair	Mean	0.211	0.214	0.230	0.225	0.238	0.217	0.232	0.242	0.229	0.236	0.249	0.260	0.221	0.251
		StdErr	0.014	0.013	0.013	0.014	0.014	0.013	0.012	0.016	0.014	0.016	0.017	0.016	0.012	0.015
		Median	0.210	0.214	0.229	0.225	0.238	0.216	0.233	0.241	0.229	0.235	0.249	0.260	0.221	0.251
		z	------	-0.172	1.231	-0.789	-1.655	-0.353	-1.360	-1.851	-1.242	1.426	2.083	2.845	-0.669	-2.241
		p	------	0.864	0.218	0.430	0.098	0.724	0.174	0.064	0.214	0.154	0.037	0.004	0.504	0.025
	Flat to Rolled, Non-Match Pair	Mean	0.236	0.216	0.243	0.235	0.251	0.244	0.221	0.249	0.245	0.229	0.255	0.231	0.253	0.276
		StdErr	0.014	0.014	0.016	0.016	0.016	0.014	0.015	0.015	0.014	0.015	0.014	0.013	0.016	0.015
		Median	0.236	0.216	0.243	0.235	0.251	0.244	0.221	0.248	0.245	0.229	0.255	0.231	0.253	0.275
		z	------	1.305	-0.432	0.055	-0.893	-0.482	0.978	-0.766	-0.639	0.430	-1.231	0.318	-1.004	-2.387
		p	------	0.192	0.666	0.956	0.372	0.630	0.328	0.444	0.523	0.667	0.218	0.750	0.315	0.017
	Flat to Flat, Match Pair	Mean	0.036	0.039	0.042	0.046	0.057	0.056	0.068	0.087	0.092	0.082	0.109	0.111	0.121	0.111
		StdErr	0.007	0.006	0.008	0.008	0.009	0.009	0.008	0.010	0.009	0.010	0.010	0.010	0.011	0.011
		Median	0.036	0.039	0.041	0.046	0.058	0.056	0.069	0.088	0.091	0.081	0.109	0.111	0.121	0.110
		z	------	-0.372	1.042	-1.176	-2.315	-2.334	-3.888	-5.403	-5.889	4.693	6.695	6.986	-7.553	-6.325
		p	------	0.710	0.298	0.240	0.021	0.020	0.000	0.000	0.000	0.000	0.000	0.000	0.000	0.000
	Flat to Flat, Non-Match Pair	Mean	0.114	0.118	0.107	0.119	0.128	0.145	0.134	0.150	0.143	0.150	0.163	0.164	0.178	0.174
		StdErr	0.011	0.011	0.010	0.011	0.010	0.012	0.011	0.012	0.012	0.012	0.012	0.011	0.012	0.013
		Median	0.114	0.118	0.108	0.119	0.129	0.145	0.134	0.150	0.143	0.150	0.163	0.164	0.179	0.174
		z	------	-0.335	0.628	-0.484	-1.295	-2.930	-2.125	-3.170	-2.867	-3.285	-4.596	-4.777	-5.413	-5.119
		p	------	0.738	0.530	0.628	0.195	0.003	0.034	0.002	0.004	0.001	0.000	0.000	0.000	0.000
	Slap to Slap, Match Pair	Mean	0.010	0.006	0.007	0.015	0.037	0.062	0.104	0.134	0.155	0.169	0.179	0.186	0.205	0.214
		StdErr	0.003	0.003	0.003	0.004	0.006	0.008	0.009	0.009	0.009	0.009	0.008	0.008	0.009	0.008
		Median	0.010	0.006	0.008	0.015	0.038	0.063	0.104	0.134	0.155	0.169	0.179	0.186	0.205	0.214
		z	------	1.028	0.659	-0.980	-3.674	-6.775	-10.188	-12.885	-16.797	-17.164	-19.494	-20.243	-21.205	-23.585
		p	------	0.304	0.510	0.327	0.000	0.000	0.000	0.000	0.000	0.000	0.000	0.000	0.000	0.000
	Slap to Slap, Non-Match Pair	Mean	0.052	0.049	0.047	0.055	0.090	0.105	0.160	0.165	0.185	0.188	0.202	0.211	0.226	0.232
		StdErr	0.008	0.007	0.007	0.007	0.009	0.009	0.009	0.009	0.008	0.008	0.008	0.008	0.007	0.009
		Median	0.053	0.049	0.048	0.055	0.090	0.105	0.160	0.165	0.185	0.188	0.203	0.211	0.226	0.233
		z	------	0.367	0.501	-0.234	-3.413	-4.710	-9.279	-9.762	-12.102	-11.890	-12.964	-14.121	-16.853	-16.904
		p	------	0.714	0.616	0.815	0.001	0.000	0.000	0.000	0.000	0.000	0.000	0.000	0.000	0.000

www.ingramcontent.com/pod-product-compliance
Lightning Source LLC
Chambersburg PA
CBHW081736170526
45167CB00009B/3844